Non-commutative Differentiation
and the
Commutator

The Search for the Fermion Content of the Universe

By

Dennis Morris

Published by: Abane & Right

56 Coach Road

Brotton

Saltburn

TS12 2RP

01287 678918

January 2018

Revised March 2018

Contents

Contents

Contents

Finite Groups to order 15:

If a finite group appears alone, as C_5, then there is only one subgroup of that form within the given finite group. Where there are multiple subgroups of the same form, they are shown like $3(C_2)$.

Order	Group	Proper subgroups
1 – Cyclic	C_1	-
2 – Cyclic	$C_2 \cong S_2$	-
3	C_3	-
4	C_4	C_2
4 – Crossed	$C_2 \times C_2$	$3(C_2)$
5	C_5	-
6	C_6	C_2, C_3
6 - Symmetric	$S_3 \cong D_3$	$3(C_2)$, C_3
7	C_7	-
8	C_8	C_2, C_4
8	$C_2 \times C_4$	$3(C_2)$, $2(C_4)$, $C_2 \times C_2$
8	$C_2 \times C_2 \times C_2$	$7(C_2)$, $7(C_2 \times C_2)$
8 – Dihedral	D_4	C_4, $5(C_2)$, $2(C_2 \times C_2)$
8 – Dicyclic	$Q \cong Q_8$	C_2, $3(C_4)$
9	C_9	C_3
9	$C_3 \times C_3$	$4(C_3)$
10	C_{10}	C_2, C_5
10 – Dihedral	D_5	$5(C_2)$, C_5
11	C_{11}	-
12	C_{12}	C_2, C_3, C_4, C_6
12	$C_2 \times C_6$	$3(C_2)$, C_3, $C_2 \times C_2$, $3(C_6)$
12 – Alternating	A_4	$3(C_2)$, $C_2 \times C_2$, $4(C_3)$
12	D_6	$7(C_2)$, C_3, C_6, $3(C_2 \times C_2)$, $2(D_3)$
12 – Dicyclic	Q_{12}	C_2, C_3, $3(C_4)$, C_6
13	C_{13}	-
14	C_{14}	C_2, C_7
14	D_7	$7(C_2)$, C_7
15	C_{15}	C_3, C_5

Preliminaries

Chapter 1

Introduction

The mathematical operation of differentiation has been known since Isaac Newton (1643-1727) and Gottfried Wilhelm Leibniz (1646-1716). Differentiation, and calculus in general, works. In millions, probably billions, of calculations, differentiation has given the correct result as verified by experiment or as verified by practical experience.

Differentiation from first principles:

Let us differentiate a simple polynomial with real variables from first principles as we did at infant school. We have:

$$y = x^2 + 3x + 4 \tag{1.1}$$

We let the variable x change by a small amount, dx. There will be a corresponding change in the y variable, dy. We now have:

$$y + dy = (x + dx)^2 + 3(x + dx) + 4 \tag{1.2}$$

Using, (1.1), we get:

$$dy = 2xdx + dx^2 + 3dx \tag{1.3}$$

We are now ready to form the differential. We divide throughout by dx. We get:

$$\frac{dy}{dx} = 2x + 3 + dx \tag{1.4}$$

We now argue that, as $dx \rightarrow 0$ so $dy \rightarrow 0$, but $\dfrac{dy}{dx}$ approaches a particular real number. Can you imagine any mathematician accepting this? We all accept this because it works – reality is the arbiter of truth. Accepting $dx \rightarrow 0$ allows us to discard the dx term on the right-hand side of (1.4) to give the differential:

$$\frac{dy}{dx} = 2x + 3 \tag{1.5}$$

Great, we know how to differentiate within the real numbers. This is all standard stuff, but there is a subtlety which we did not mention above. We could unambiguously form the differential because we were working in the real numbers and the real numbers are a multiplicatively commutative division algebra.

Differentiation within non-commutative algebras:

Now suppose we are working in a multiplicatively non-commutative division algebra such as the quaternions. We have a problem:

$$dy\frac{1}{dx} \neq \frac{1}{dx}dy \tag{1.6}$$

Whoops! We cannot unambiguously form the usual differential in a non-commutative algebra. How can we form the differential within a non-commutative algebra?

This book presents the theory of non-commutative differentiation.

We will see that we need to go to the very basics of mathematics to develop non-commutative differentiation. We cannot quickly pull a non-commutative rabbit out of a hat and expect the reader to applaud. We must base our non-commutative differentiation upon solid foundations.

We will see that our derivation of non-commutative differentiation leads us down paths previously untrodden and along paths that were trodden in ancient times but are now overgrown and forgotten. Remarkably, these paths lead directly to the physical universe.

Think of this book as an adventure story.

Chapter 2

Finite Group Representations

Well! you might think that finite groups cannot possibly have anything to do with differentiation. You are wrong. In this chapter, we acquaint the reader with aspects of finite groups which were once known but have now been largely forgotten.

Permutations – our first step:

A finite group is a closed set of permutations. We need no more that this; group actions, category theory, we do not need it. We will be dealing with nothing more than closed sets of permutations.

We can write any finite group as a set of square[1] permutation matrices of size equal to the order of the group using only zero and plus-one. For example, the finite cyclic group of order two, C_2, is represented by the two 2×2 permutation matrices below, (2.1), and the finite cyclic group of order three, C_3, is represented by the three permutation matrices below, (2.1):

$$C_2 = \left\{ \begin{bmatrix} 1 & 0 \\ 0 & 1 \end{bmatrix}, \begin{bmatrix} 0 & 1 \\ 1 & 0 \end{bmatrix} \right\} \qquad C_3 = \left\{ \begin{bmatrix} 1 & 0 & 0 \\ 0 & 1 & 0 \\ 0 & 0 & 1 \end{bmatrix}, \begin{bmatrix} 0 & 1 & 0 \\ 0 & 0 & 1 \\ 1 & 0 & 0 \end{bmatrix}, \begin{bmatrix} 0 & 0 & 1 \\ 1 & 0 & 0 \\ 0 & 1 & 0 \end{bmatrix} \right\} \qquad (2.1)$$

Notice that there are no minus ones in the above representations, (2.1), and that the matrices are of the same size as the number of elements, matrices, in the group. The number of elements in a finite group is called the order of the group.

Instead of thinking of a finite group as a closed set of permutations, the reader can think of a finite group as a closed set of permutation matrices. By closed, we mean closed under matrix multiplication.

Another step:

Some finite groups can be written as sets of permutation matrices using matrices of half the size of the order of the group if we use minus-one together with zero and plus-one; for example, the finite cyclic group of order four, C_4, is represented by the four 2×2 permutation matrices given just below, (2.2):

$$C_4 = \left\{ \begin{bmatrix} 1 & 0 \\ 0 & 1 \end{bmatrix}, \begin{bmatrix} 0 & -1 \\ 1 & 0 \end{bmatrix}, \begin{bmatrix} -1 & 0 \\ 0 & -1 \end{bmatrix}, \begin{bmatrix} 0 & 1 \\ -1 & 0 \end{bmatrix} \right\} \qquad (2.2)$$

---

[1] All permutation matrices are square.

There is often more than one set of permutation matrices which represent a given group. The finite cyclic group of order four can be written in three ways using 4×4 matrices:

$$C_4 = \left\{ \begin{bmatrix} 1 & 0 & 0 & 0 \\ 0 & 1 & 0 & 0 \\ 0 & 0 & 1 & 0 \\ 0 & 0 & 0 & 1 \end{bmatrix}, \begin{bmatrix} 0 & 1 & 0 & 0 \\ 0 & 0 & 1 & 0 \\ 0 & 0 & 0 & 1 \\ 1 & 0 & 0 & 0 \end{bmatrix}, \begin{bmatrix} 0 & 0 & 1 & 0 \\ 0 & 0 & 0 & 1 \\ 1 & 0 & 0 & 0 \\ 0 & 1 & 0 & 0 \end{bmatrix}, \begin{bmatrix} 0 & 0 & 0 & 1 \\ 1 & 0 & 0 & 0 \\ 0 & 1 & 0 & 0 \\ 0 & 0 & 1 & 0 \end{bmatrix} \right\}$$

(2.3)

A Standard Representation of Cyclic groups

This first representation of C_4, (2.3), with all the ones running parallel to the leading diagonal is a standard way of representing a cyclic group. Any cyclic group of any order has a representation with lines of ones running 'parallel' to the leading diagonal like the above, (2.3).

We also have:

$$C_4 = \left\{ \begin{bmatrix} 1 & 0 & 0 & 0 \\ 0 & 1 & 0 & 0 \\ 0 & 0 & 1 & 0 \\ 0 & 0 & 0 & 1 \end{bmatrix}, \begin{bmatrix} 0 & 1 & 0 & 0 \\ 1 & 0 & 0 & 0 \\ 0 & 0 & 0 & 1 \\ 0 & 0 & 1 & 0 \end{bmatrix}, \begin{bmatrix} 0 & 0 & 1 & 0 \\ 0 & 0 & 0 & 1 \\ 0 & 1 & 0 & 0 \\ 1 & 0 & 0 & 0 \end{bmatrix}, \begin{bmatrix} 0 & 0 & 0 & 1 \\ 0 & 0 & 1 & 0 \\ 1 & 0 & 0 & 0 \\ 0 & 1 & 0 & 0 \end{bmatrix} \right\}$$

(2.4)

And:

$$C_4 = \left\{ \begin{bmatrix} 1 & 0 & 0 & 0 \\ 0 & 1 & 0 & 0 \\ 0 & 0 & 1 & 0 \\ 0 & 0 & 0 & 1 \end{bmatrix}, \begin{bmatrix} 0 & 1 & 0 & 0 \\ 0 & 0 & 0 & 1 \\ 1 & 0 & 0 & 0 \\ 0 & 0 & 1 & 0 \end{bmatrix}, \begin{bmatrix} 0 & 0 & 1 & 0 \\ 1 & 0 & 0 & 0 \\ 0 & 0 & 0 & 1 \\ 0 & 1 & 0 & 0 \end{bmatrix}, \begin{bmatrix} 0 & 0 & 0 & 1 \\ 0 & 0 & 1 & 0 \\ 0 & 1 & 0 & 0 \\ 1 & 0 & 0 & 0 \end{bmatrix} \right\}$$

(2.5)

Another example of multiple finite group representations of the same size is the two 4-dimensional representations of the order eight quaternion finite group:

$$Q_8^{L\chi} = \left\{ \begin{array}{l} \begin{bmatrix} 1 & 0 & 0 & 0 \\ 0 & 1 & 0 & 0 \\ 0 & 0 & 1 & 0 \\ 0 & 0 & 0 & 1 \end{bmatrix}, \begin{bmatrix} 0 & 1 & 0 & 0 \\ -1 & 0 & 0 & 0 \\ 0 & 0 & 0 & -1 \\ 0 & 0 & 1 & 0 \end{bmatrix}, \begin{bmatrix} 0 & 0 & 1 & 0 \\ 0 & 0 & 0 & 1 \\ -1 & 0 & 0 & 0 \\ 0 & -1 & 0 & 0 \end{bmatrix}, \begin{bmatrix} 0 & 0 & 0 & 1 \\ 0 & 0 & -1 & 0 \\ 0 & 1 & 0 & 0 \\ -1 & 0 & 0 & 0 \end{bmatrix}, \\[30pt] \begin{bmatrix} -1 & 0 & 0 & 0 \\ 0 & -1 & 0 & 0 \\ 0 & 0 & -1 & 0 \\ 0 & 0 & 0 & -1 \end{bmatrix}, \begin{bmatrix} 0 & -1 & 0 & 0 \\ 1 & 0 & 0 & 0 \\ 0 & 0 & 0 & 1 \\ 0 & 0 & -1 & 0 \end{bmatrix}, \begin{bmatrix} 0 & 0 & -1 & 0 \\ 0 & 0 & 0 & -1 \\ 1 & 0 & 0 & 0 \\ 0 & 1 & 0 & 0 \end{bmatrix}, \begin{bmatrix} 0 & 0 & 0 & -1 \\ 0 & 0 & 1 & 0 \\ 0 & -1 & 0 & 0 \\ 1 & 0 & 0 & 0 \end{bmatrix} \end{array} \right\}$$

(2.6)

And:

$$Q_8^{R\chi} = \left\{ \begin{bmatrix} 1 & 0 & 0 & 0 \\ 0 & 1 & 0 & 0 \\ 0 & 0 & 1 & 0 \\ 0 & 0 & 0 & 1 \end{bmatrix}, \begin{bmatrix} 0 & 1 & 0 & 0 \\ -1 & 0 & 0 & 0 \\ 0 & 0 & 0 & 1 \\ 0 & 0 & -1 & 0 \end{bmatrix}, \begin{bmatrix} 0 & 0 & 1 & 0 \\ 0 & 0 & 0 & -1 \\ -1 & 0 & 0 & 0 \\ 0 & 1 & 0 & 0 \end{bmatrix}, \begin{bmatrix} 0 & 0 & 0 & 1 \\ 0 & 0 & 1 & 0 \\ 0 & -1 & 0 & 0 \\ -1 & 0 & 0 & 0 \end{bmatrix}, \\ \begin{bmatrix} -1 & 0 & 0 & 0 \\ 0 & -1 & 0 & 0 \\ 0 & 0 & -1 & 0 \\ 0 & 0 & 0 & -1 \end{bmatrix}, \begin{bmatrix} 0 & -1 & 0 & 0 \\ 1 & 0 & 0 & 0 \\ 0 & 0 & 0 & -1 \\ 0 & 0 & 1 & 0 \end{bmatrix}, \begin{bmatrix} 0 & 0 & -1 & 0 \\ 0 & 0 & 0 & 1 \\ 1 & 0 & 0 & 0 \\ 0 & -1 & 0 & 0 \end{bmatrix}, \begin{bmatrix} 0 & 0 & 0 & -1 \\ 0 & 0 & -1 & 0 \\ 0 & 1 & 0 & 0 \\ 1 & 0 & 0 & 0 \end{bmatrix} \right\} \quad (2.7)$$

Equivalent representations?

Any set of matrices can be rewritten in a different basis. If we do this with a finite group representation such as (2.3) or (2.7), then we still have a representation of the original finite group, but such representations are all equivalent. They look less tidy. We have no interest in such equivalent representations.

Inequivalent representations:

The three representations of the finite group C_4, (2.3) & (2.4) & (2.5) differ from each other by a swapping of two variables – one variable is unchanged.

We will not consider these three representations of the finite group C_4, (2.3) & (2.4) & (2.5), to be equivalent. We will deal with each such representation as existing in its own right. This might seem overly pedantic and perhaps a little shocking to the reader, but we will soon see that this is a great step forward for humankind.

Take a look at the two representations of the quaternion group, (2.6) & (2.7). The reader might notice that there are different super-scripts on the Q_8. The commutation relations of these two representations, (2.6) & (2.7), are opposite to each other. The superscripts $R\chi$ & $L\chi$ indicate that one representation is right-handed, right-chiral, and the other representation is left-handed, left-chiral. To view these two quaternion group representations, (2.6) & (2.7), as equivalent is to miss this chiral subtlety; we do not walk this path.

An aside - Chirality:

We emphasize that the chirality of a representation of a finite group is in the commutation relations.

Chirality will be discussed at much greater length later in this book, and it will play a central role in non-commutative differentiation. Do not worry if you know nothing of chirality; things will be made much clearer in later chapters.

Summary:

We are going to form two types of representations of the finite groups. One of these finite group representations will be square matrices of the same size as the order of the group and containing only zeros and plus-ones. The other of these finite group representations will be square matrices of the half the size as the order of the group and will containing only zeros and plus-ones and minus-ones. It will transpire that our primary interest is in the second type, size equal to half the group order, of these two types of finite group representations.

Chapter 3

Finite Group Spaces

We will now look at work done in ancient times, a hundred years ago, by Ferdinand Georg Frobenius (1849-1917) and Richard Dedekind (1831-1916) and since then forgotten.

The finite group space:

We take a representation of a finite group written as a number of matrices – one matrix for each element of the particular finite group. We multiply each of these matrices by a real number; we never use complex numbers within a matrix – you will see why shortly. We add the so modified matrices of the representation, and we take the exponential of this sum of modified matrices.

For example, we choose the order two cyclic finite group C_2, (2.1). Firstly, multiply each matrix of the representation by a different real variable:

$$C_2 = \left\{ \begin{bmatrix} 1 & 0 \\ 0 & 1 \end{bmatrix}, \begin{bmatrix} 0 & 1 \\ 1 & 0 \end{bmatrix} \right\} \rightarrow \begin{bmatrix} t & 0 \\ 0 & t \end{bmatrix}, \begin{bmatrix} 0 & z \\ z & 0 \end{bmatrix} \quad : \quad \{t, z\} \in \mathbb{R} \tag{3.1}$$

Secondly, add the matrices:

$$\begin{bmatrix} t & 0 \\ 0 & t \end{bmatrix} + \begin{bmatrix} 0 & z \\ z & 0 \end{bmatrix} = \begin{bmatrix} t & z \\ z & t \end{bmatrix}$$

The algebraic matrix form $\qquad\qquad$ (3.2)

The representation module

The representation space

The resulting sum of the matrices is called the algebraic matrix form. In the past, it has been called the representation module or the representation space[2].

Thirdly, we take the exponential of the algebraic matrix form:

$$\exp\left(\begin{bmatrix} t & z \\ z & t \end{bmatrix}\right) = \begin{bmatrix} e^t & 0 \\ 0 & e^t \end{bmatrix} \begin{bmatrix} \cosh z & \sinh z \\ \sinh z & \cosh z \end{bmatrix} = \begin{bmatrix} r & 0 \\ 0 & r \end{bmatrix} \begin{bmatrix} \cosh z & \sinh z \\ \sinh z & \cosh z \end{bmatrix} \tag{3.3}$$

We see that the Lorentz transformation, the Lorentz velocity boost, of 2-dimensional space-time has appeared out of the finite group C_2. We say that 2-dimensional space-time is a finite group space – it is a geometric space that has emerged from a finite group.

---

[2] These names were coined by Dedekind and Frobenius.

The Lorentz velocity boost is a rotation in 2-dimensional space-time. It corresponds to a change of velocity – hence the name 'boost'. However, we must emphasize that a Lorentz boost is a rotation. A change of velocity is no more than a rotation in 2-dimensional space-time. The matrix above containing the hyperbolic trigonometric functions, (3.3), cosh and sinh, is a rotation matrix.

An aside: Orthogonality and rotation matrices:
We know the matrix in (3.3) is not an orthogonal matrix. We know there are rumours that all rotation matrices are orthogonal. We utterly reject the idea that a matrix containing trigonometric functions is not a rotation matrix because it is not orthogonal – that nonsense wants to be put straight in the bin. True, all Euclidean rotations are orthogonal matrices, but we are not restricted to Euclidean space.

Back to finite group spaces:
In (3.3), we have seen a rotation matrix emerge from a representation of a finite group.

Clearly the exponential form is equal to the Cartesian form for some values of the variables:

$$\begin{bmatrix} r & 0 \\ 0 & r \end{bmatrix}\begin{bmatrix} \cosh\varphi & \sinh\varphi \\ \sinh\varphi & \cosh\varphi \end{bmatrix} = \begin{bmatrix} t & z \\ z & t \end{bmatrix} \tag{3.4}$$

Now, the rotation matrix is formed as the exponential of a matrix with zero trace. It is a basic mathematical identity that the determinant of the exponential of a matrix with zero trace will always be unity. Thus, all rotation matrices have determinant unity.

We take the determinant of both sides of (3.4), and we get:

$$\det\left(\begin{bmatrix} r & 0 \\ 0 & r \end{bmatrix}\begin{bmatrix} \cosh\varphi & \sinh\varphi \\ \sinh\varphi & \cosh\varphi \end{bmatrix}\right) = \det\left(\begin{bmatrix} t & z \\ z & t \end{bmatrix}\right) \tag{3.5}$$

$$r^2 = t^2 - z^2$$

We have the distance function of 2-dimensional space-time[3]. We are now in a position to state:

$$\text{The Finite Group Space of the } 2\times 2 \text{ representation of the finite} \atop \text{group } C_2 \text{ is 2-dimensional space-time.} \tag{3.6}$$

We observe 2-dimensional space-time within our 4-dimensional space-time. It is a physically real space. What a jolly exciting adventure we are having.

---

[3] Frobenius used to refer to the determinant of the algebraic matrix form as the 'Group Determinant', but this terminology seems to have been lost in history.

An aside: The special theory of relativity:

The theory of special relativity is no more than the statement that the laws of physics are invariant under a rotation in 2-dimensional space-time[4] - a kettle will boil at the same temperature regardless of the velocity at which the kettle is moving[5].

In general, physics is invariant under rotation in any kind of finite group space. A car engine works exactly the same when the car is heading northward as when the car is heading eastward. Kettles boil at the same temperature when the spout is pointing westward as when the spout is pointing southward. This is just another way of saying the finite group spaces are isotropic (the same in all directions).

Let's do it again:

We see that we can produce a geometric space with a rotation matrix, including trigonometric functions, and a distance function from a representation of a finite group. How about the 2-dimensional representation of the order four cyclic finite group given above, (2.2)? We have:

$$C_4 = \left\{ \begin{bmatrix} 1 & 0 \\ 0 & 1 \end{bmatrix}, \begin{bmatrix} 0 & -1 \\ 1 & 0 \end{bmatrix}, \begin{bmatrix} -1 & 0 \\ 0 & -1 \end{bmatrix}, \begin{bmatrix} 0 & 1 \\ -1 & 0 \end{bmatrix} \right\} \tag{3.7}$$

Multiply each matrix by a real number and add these matrices:

$$\begin{bmatrix} a & 0 \\ 0 & a \end{bmatrix} + \begin{bmatrix} 0 & -b \\ b & 0 \end{bmatrix} + \begin{bmatrix} -c & 0 \\ 0 & -c \end{bmatrix} + \begin{bmatrix} 0 & d \\ -d & 0 \end{bmatrix} = \begin{bmatrix} a-c & d-b \\ -(d-b) & a-c \end{bmatrix} = \begin{bmatrix} x & y \\ -y & x \end{bmatrix} \tag{3.8}$$

Take the exponential:

$$\exp\left(\begin{bmatrix} x & y \\ -y & x \end{bmatrix} \right) = \begin{bmatrix} r & 0 \\ 0 & r \end{bmatrix} \begin{bmatrix} \cos y & \sin y \\ -\sin y & \cos y \end{bmatrix} \tag{3.9}$$

And the determinant is:

$$\det\left(\begin{bmatrix} r & 0 \\ 0 & r \end{bmatrix} \begin{bmatrix} \cos y & \sin y \\ -\sin y & \cos y \end{bmatrix} \right) = \det\left(\begin{bmatrix} a & b \\ -b & a \end{bmatrix} \right) \tag{3.10}$$

$$r^2 = a^2 + b^2$$

We now have 2-dimensional Euclidean space emerging from the 2-dimensional representation of the order four finite cyclic group C_4. We have the rotation matrix of 2-dimensional Euclidean space together with the Euclidean 2-dimensional trigonometric functions. We have the 2-dimensional Euclidean distance function.

---

[4] Many authors have tried to present special relativity as a 4-dimensional theory using 4-vectors. We opine that this is nonsense and that special relativity is a 2-dimensional theory.
[5] This 'kettle' thing really is the special theory of relativity in a tea-cup. The velocity of light determines the strength of the inter-molecular forces between the water molecules and hence determines the boiling point of water.

"What more do you want?"

"We want a divergence and a curl".

"Don't be impatient. That will come later".

The adventure is getting even jollier.

Division algebras:

Consider the complex number product (written in ancient notation[6]):

$$(a+\hat{i}b)(c+\hat{i}d) = ac - bd + \hat{i}(ad+bc) \tag{3.11}$$

Consider also the complex number product (written in modern notation):

$$\begin{bmatrix} a & b \\ -b & a \end{bmatrix}\begin{bmatrix} c & d \\ -d & c \end{bmatrix} = \begin{bmatrix} ac-bd & ad+bc \\ -(ad+bc) & ac-bd \end{bmatrix} \tag{3.12}$$

We will abandon the ancient notation for complex numbers; it obscures so much. From now onward, we will write a complex number as a 2×2 matrix. Above, (3.9) & (3.10) is a Euclidean complex number, $\mathbb{C}$.

We have seen a kind of numbers, a division algebra, the 2-dimensional complex numbers, $\mathbb{C}$, arise from the order four cyclic finite group. Associated with this kind of numbers is a kind of geometric space with a rotation matrix and a distance function held invariant by that rotation matrix. We call this geometric space the complex plane or simply 2-dimensional Euclidean space, but we must be careful. The complex plane is not the same as $\mathbb{R}^2$. The complex plane has an imaginary axis and a single real axis. $\mathbb{R}^2$ has two real axes[7].

The complex plane is a finite group space. It is a geometric space which emerges from a representation of a finite group. It is a division algebra which emerges from a finite group. In general, every finite group space is a kind of numbers; every finite group space is a kind of division algebra.

The 2-dimensional space-time that we derived from the order two finite cyclic group C_2, above, (3.5), is the hyperbolic complex numbers, $\mathbb{S}$, first discovered by James Cockle (1819-1895) in 1848[8].

3-dimensional rotations:

The reader will be observationally familiar with only the two types of 2-dimensional rotations we have derived above, rotation in 2-dimensional space-time – a velocity boost, (3.3), and rotation in 2-dimensional Euclidean space, (3.9). We are familiar with these two kinds of rotation because the distance function of our 4-dimensional space-time can accommodate the distance functions of both these two types of 2-dimensional rotations.

---

[6] Okay, this notation is not ancient at all, but it will be one day; we are just forward looking.

[7] We are not convinced the $\mathbb{R}^2$ really exists.

[8] Cockle called the hyperbolic complex numbers 'Tessarines'.

Higher dimensional rotation:

Although it might be a conceptual shock to the reader, there is no reason why rotations cannot be of dimension higher than two.

We take the 3-dimensional representation of the order three finite group, C_3, and we form the finite group space of this finite group:

$$C_3 = \left\{ \begin{bmatrix} 1 & 0 & 0 \\ 0 & 1 & 0 \\ 0 & 0 & 1 \end{bmatrix}, \begin{bmatrix} 0 & 1 & 0 \\ 0 & 0 & 1 \\ 1 & 0 & 0 \end{bmatrix}, \begin{bmatrix} 0 & 0 & 1 \\ 1 & 0 & 0 \\ 0 & 1 & 0 \end{bmatrix} \right\}$$

(3.13)

$$\begin{bmatrix} a & 0 & 0 \\ 0 & a & 0 \\ 0 & 0 & a \end{bmatrix} + \begin{bmatrix} 0 & b & 0 \\ 0 & 0 & b \\ b & 0 & 0 \end{bmatrix} + \begin{bmatrix} 0 & 0 & c \\ c & 0 & 0 \\ 0 & c & 0 \end{bmatrix} = \begin{bmatrix} a & b & c \\ c & a & b \\ b & c & a \end{bmatrix}$$

Taking the exponential gives:

$$\exp\left(\begin{bmatrix} a & b & c \\ c & a & b \\ b & c & a \end{bmatrix} \right) = \begin{bmatrix} e^a & 0 & 0 \\ 0 & e^a & 0 \\ 0 & 0 & e^a \end{bmatrix} \begin{bmatrix} v_A(b,c) & v_B(b,c) & v_C(b,c) \\ v_C(b,c) & v_A(b,c) & v_B(b,c) \\ v_B(b,c) & v_C(b,c) & v_A(b,c) \end{bmatrix}$$

(3.14)

The matrix with the nu-functions, $v_i(b,c)$, is a 3-dimensional rotation matrix. The nu-functions, $v_i(b,c)$, are 3-dimensional trigonometric functions[9]. Notice that a 3-dimensional angle is comprised of two variables[10]. This is a general phenomenon; n-dimensional rotations are associated the angles comprised of $(n-1)$ real variables.

All rotations hold invariant the distance from the centre of rotation to a moving point. However different types of rotation hold invariant different types of distance. What kind of distance does this 3-dimensional rotation hold invariant? Simple! Just take the determinant of the algebraic matrix form:

$$\det\left(\begin{bmatrix} r & 0 & 0 \\ 0 & r & 0 \\ 0 & 0 & r \end{bmatrix} \begin{bmatrix} v_A(b,c) & v_B(b,c) & v_C(b,c) \\ v_C(b,c) & v_A(b,c) & v_B(b,c) \\ v_B(b,c) & v_C(b,c) & v_A(b,c) \end{bmatrix} \right) = \det\left(\begin{bmatrix} a & b & c \\ c & a & b \\ b & c & a \end{bmatrix} \right)$$

(3.15)

$$r^3 = a^3 + b^3 + c^3 - 3abc$$

This, (3.15), is the distance function of this 3-dimensional finite group space; it is the distance held invariant by this 3-dimensional rotation, (3.15).

---

[9] These 3-dimensional trigonometric functions are discussed at length in the book : Dennis Morris : Complex Numbers The Higher Dimensional Forms. They are hyper-geometric functions. These are very interesting functions, but they are quite unexplored.
[10] These are real variables. We use only real variables in our matrices.

What is a rotation?

We often see a rotation to be defined as the set of orthogonal matrices. If we were to restrict ourselves to only 2-dimensional Euclidean space, this would be true; but we do not wish to restrict ourselves to only one kind of space.

We might define a rotation matrix, and hence a rotation, as a matrix full of functions of variables such that, as these variables vary from minus infinity to plus infinity, the matrix holds invariant a particular expression which we call the distance function. This is the idea that a rotation moves a point in such a way as to keep constant the distance, as measured by the distance function of the space, from the centre of rotation – a rotation traces out the set of points equidistant from the origin - a circle in a most general sense.

We might define a rotation as being the locus of a point such that the position on one axis is equal to the absolute rate of change of position on another axis. This is the differential relations between the trigonometric[11] functions in the rotation matrix:

$$\frac{\partial \sin x}{\partial x} = \cos x \qquad \frac{\partial \cos x}{\partial x} = -\sin x \qquad : \qquad \frac{\partial \sinh x}{\partial x} = \cosh x \qquad \frac{\partial \cosh x}{\partial x} = \sinh x$$

$$(3.16)$$

This says effectively that the trigonometric functions within the rotation matrix are closely related to the exponential series.

We might define the trigonometric functions as being projections from the unit 'circle' on to each axis. This is how the 2-dimensional Euclidean trigonometric functions were first defined in 5th century India[12].

How-so-ever we define a rotation other than the very restrictive orthogonal way, the above matrix of nu-functions, (3.15), satisfies that definition. Indeed, every rotation matrix we can derive as part of a finite group space will satisfy that definition and every nu-function within those rotation matrices will be a trigonometric function. Do not try to tell me that the 3×3 matrix above, (3.14), containing the nu-functions is not a rotation matrix just because you've never seen this kind of rotation before.

Why we do not see 3-dimensional rotations in our 4-dimensional space-time:

The distance function of our 4-dimensional space-time is:

$$dist^2 = t^2 - x^2 - y^2 - z^2 \qquad (3.17)$$

If we set any two variables in the 4-dimensional space-time distance function, (3.17), to zero, we get the distance function of one or other of the 2-dimensional finite group spaces we derived above, (3.5) & (3.10). However, if we set any one of the variables in the 4-dimensional space-time distance function, (3.17), to zero, we do not get the distance function of the 3-dimensional finite group space we derived above, (3.15). The distance function of our 4-dimensional space-time cannot accommodate the 3-dimensional distance function and so our 4-dimensional space-time cannot

---

[11] The word trigonometry derives from the Greek word *trigonon* meaning triangle.
[12] The Aryabhata.

accommodate the 3-dimensional rotations which hold invariant the 3-dimensional distance function. That is why we never see 3-dimensional rotations in our 4-dimensional space-time.

Suppose our 4-dimensional space-time had the distance function:

$$dist^3 = a^3 + b^3 + c^3 + d^3 - 3abc - 3abd - 3acd - 3bcd \tag{3.18}$$

We see that, if we were to set one variable to zero in this proposed distance function, (3.18), then we would get a distance function kept invariant by the 3-dimensional rotation above, (3.14). This proposed distance function, (3.18), would hold four 3-dimensional rotations, but it would not hold any 2-dimensional rotations.

In due course, we will see that the above proposed distance function, (3.18), would hold the 3-dimensional curl of the 3-dimensional finite group space of the order three cyclic finite group C_3, but that our 4-dimensional space-time distance function, (3.17), will not hold any 3-dimensional curl just as it will not hold any 3-dimensional rotation.

Did you get that last bit? The only types of curls that can be manifest in our 4-dimensional space-time are those that are of the same form as the curls that are associated with the two kinds of 2-dimensional rotation. This means the only fields that can be manifest in our 4-dimensional space-time are fields with curls like the two 2-dimensional curls – a small number of fields – a small number of types of force.

Summary:
In this chapter, we have introduced the reader to the finite group spaces. A finite group space consists of a distance function and a rotation matrix which respects that distance function. Within the rotation matrix are the trigonometric functions which vary in such a way as to hold the distance function invariant for different values of the variables, angle, that are the argument of those trigonometric functions.

We have seen geometric spaces like the complex plane emerge from representations of finite groups. Each of these geometric spaces is called a finite group space. Every finite group space is a division algebra[13] like the 2-dimensional complex numbers, $\mathbb{C}$.

We need a multiplication operation to be able to form a derivative. We have seen that the multiplication operation within the 2-dimensional Euclidean complex numbers is really just matrix multiplication. Indeed, it is no more than the multiplication of permutation matrices, and that is no more than the sequential combination of permutations. We will look at algebraic operations like multiplication in more detail later – we need a non-commutative algebraic operation to form a non-commutative derivative.

We now have a fundamental understanding of finite group spaces.

---

[13] We will need to qualify this later.

Chapter 4

Quaternions and Rotation

In the first chapter of this book, we wrote of differentiation within a non-commutative algebra. To paraphrase Mrs Beeton, first catch your non-commutative algebra. We begin with the quaternions.

A non-commutative rotation:

Let us form a 4-dimensional quaternion finite group space. We begin with the right-chiral 4-dimensional representation given above, (2.7).

$$
Q_8^{R\chi} = \left\{
\begin{bmatrix} 1 & 0 & 0 & 0 \\ 0 & 1 & 0 & 0 \\ 0 & 0 & 1 & 0 \\ 0 & 0 & 0 & 1 \end{bmatrix},
\begin{bmatrix} 0 & 1 & 0 & 0 \\ -1 & 0 & 0 & 0 \\ 0 & 0 & 0 & 1 \\ 0 & 0 & -1 & 0 \end{bmatrix},
\begin{bmatrix} 0 & 0 & 1 & 0 \\ 0 & 0 & 0 & -1 \\ -1 & 0 & 0 & 0 \\ 0 & 1 & 0 & 0 \end{bmatrix},
\begin{bmatrix} 0 & 0 & 0 & 1 \\ 0 & 0 & 1 & 0 \\ 0 & -1 & 0 & 0 \\ -1 & 0 & 0 & 0 \end{bmatrix},
\right.
$$
$$
\left.
\begin{bmatrix} -1 & 0 & 0 & 0 \\ 0 & -1 & 0 & 0 \\ 0 & 0 & -1 & 0 \\ 0 & 0 & 0 & -1 \end{bmatrix},
\begin{bmatrix} 0 & -1 & 0 & 0 \\ 1 & 0 & 0 & 0 \\ 0 & 0 & 0 & -1 \\ 0 & 0 & 1 & 0 \end{bmatrix},
\begin{bmatrix} 0 & 0 & -1 & 0 \\ 0 & 0 & 0 & 1 \\ 1 & 0 & 0 & 0 \\ 0 & -1 & 0 & 0 \end{bmatrix},
\begin{bmatrix} 0 & 0 & 0 & -1 \\ 0 & 0 & -1 & 0 \\ 0 & 1 & 0 & 0 \\ 1 & 0 & 0 & 0 \end{bmatrix}
\right\} \quad (4.1)
$$

Analogously to what we did above, (3.8), we multiply each matrix by a real number and sum the matrices to get the right-chiral quaternion algebraic matrix form:

$$
\mathbb{H}_{R\chi} = \begin{bmatrix} a & b & c & d \\ -b & a & d & -c \\ -c & -d & a & b \\ -d & c & -b & a \end{bmatrix} \quad (4.2)
$$

This, (4.2), is a right-chiral quaternion. There are two kinds of quaternion; there are the left-chiral quaternions and there are right-chiral quaternions. These quaternions seem to correspond to the left-handed electron and the right-handed electron of electro-weak theory[14].

The chirality is encoded in the distribution of minus signs within the bottom right-hand 3×3 corner of the two quaternion algebraic matrix forms. It turns out that this is closely related to the chirality of our 4-dimensional space-time.

We set the leading diagonal, the identity variable, in (4.2) to zero, and we take the exponential to get the quaternion rotation matrix:

---

[14] See: Dennis Morris : The Quaternion Dirac Equation.

$$\exp\left(\begin{bmatrix} 0 & b & c & d \\ -b & 0 & d & -c \\ -c & -d & 0 & b \\ -d & c & -b & 0 \end{bmatrix}\right) = \begin{bmatrix} \cos\lambda & \dfrac{b}{\lambda}\sin\lambda & \dfrac{c}{\lambda}\sin\lambda & \dfrac{d}{\lambda}\sin\lambda \\ -\dfrac{b}{\lambda}\sin\lambda & \cos\lambda & \dfrac{d}{\lambda}\sin\lambda & -\dfrac{c}{\lambda}\sin\lambda \\ -\dfrac{c}{\lambda}\sin\lambda & -\dfrac{d}{\lambda}\sin\lambda & \cos\lambda & \dfrac{b}{\lambda}\sin\lambda \\ -\dfrac{d}{\lambda}\sin\lambda & \dfrac{c}{\lambda}\sin\lambda & -\dfrac{b}{\lambda}\sin\lambda & \cos\lambda \end{bmatrix} \qquad (4.3)$$

$$\lambda = \sqrt{b^2 + c^2 + d^2}$$

The functions within this rotation matrix, (4.3), are the quaternion trigonometric functions. The argument, angle, of the quaternion trigonometric functions is comprised of three variables[15].

The right-chiral quaternion trigonometric functions are identical to the left-chiral quaternion trigonometric functions. The chirality is in the distribution of the minus signs within the rotation matrix and not within the trigonometric functions.

If we set the $c = d = 0$ variables to zero in (4.3), we get:

$$\mathbb{H}_{R\chi}^{Rot_{c=d=0}} = \begin{bmatrix} \cos b & \sin b & 0 & 0 \\ -\sin b & \cos b & 0 & 0 \\ 0 & 0 & \cos b & \sin b \\ 0 & 0 & -\sin b & \cos b \end{bmatrix} \qquad (4.4)$$

We see that the '2-dimensional' version of the 4-dimensional right-chiral quaternion rotation matrix has twice the rotation of a 2-dimensional Euclidean rotation matrix. We opine that this is why the electron has twice the magnetic dipole moment that we might have expected.

We have all possible kinds of rotation:
A rotation is a linear transformation and must therefore be associated with a matrix, which we call a rotation matrix. The nature of rotations is such that two rotations combine to form a third rotation. This means that the form of a rotation matrix must be multiplicatively closed. It is also the case that for every rotation there is an inverse rotation. There is always a zero rotation which is the identity matrix. In short, a rotation matrix is an infinite group. With thought we see that, given enough time, we could list every possible rotation provided we can list every possible representation of every finite group. In fact, we need only the two types of representations with which we deal in this book.

We thus have a list of all possible rotations. We will see, in due course, that this implies that we also have a list of every possible type of curl.

---

[15] We always use real variables in our matrices.

Our 4-dimensional space-time is not a finite group space:

In case you were wondering, our 4-dimensional space-time is not a finite group space. This is easy to see in two ways. The first way is to simply calculate all possible 4-dimensional finite group spaces and we see that our 4-dimensional space-time is not one of them. The second way is to consider the multiplicative closure of algebraic matrix forms; for example, we take the 2-dimensional representation of the order two finite cyclic group C_2:

$$\begin{bmatrix} t & z \\ z & t \end{bmatrix}\begin{bmatrix} s & y \\ y & s \end{bmatrix} = \begin{bmatrix} st + yz & ty + sz \\ ty + sz & st + yz \end{bmatrix} \tag{4.5}$$

We see that the form of the matrices is preserved under matrix multiplication. This is a consequence of this algebraic matrix forms being derived from a finite group. A consequence of this multiplicative closure of matrix form is the multiplicative closure of determinant form:

$$\det\left(\begin{bmatrix} t & z \\ z & t \end{bmatrix}\right)\det\left(\begin{bmatrix} s & y \\ y & s \end{bmatrix}\right) = \det\left(\begin{bmatrix} t & z \\ z & t \end{bmatrix}\begin{bmatrix} s & y \\ y & s \end{bmatrix}\right) \tag{4.6}$$

$$\left(t^2 - z^2\right)\left(s^2 - y^2\right) = \left(st + yz\right)^2 - \left(ty + sz\right)^2$$

Very simply, the distance function of our 4-dimensional space-time does not have this property of multiplicative closure of form.

An aside:

In the traditional notation, a quaternion is written as:

$$\mathbb{H} = a + \hat{i}b + jc + kd \tag{4.7}$$

There is no hint of chirality within this traditional notation, (4.7). By changing the notation, we reveal an important property of the quaternions. If we set the $c = d = 0$ variables to zero in the traditional notation, we get:

$$\mathbb{C}? = a + \hat{i}b \tag{4.8}$$

This was traditionally taken as being a 2-dimensional complex number, but it is not. With the matrices in mind, (3.9) & (4.4), we see that there is a big difference between a 2-dimensional rotation and a 4-dimensional rotation in a 2-dimensional plane – the electron knows the difference.

Why we do not use complex numbers in matrices:

We have seen above, (3.9), that a Euclidean complex number can be written as a 2×2 matrix:

$$a + \hat{i}b = \begin{bmatrix} a & b \\ -b & a \end{bmatrix} = \mathbb{C} \tag{4.9}$$

We could write a left-chiral quaternion as the 2×2 matrix:

$$\mathbb{H}_{L\chi} = \begin{bmatrix} a & b & c & d \\ -b & a & -d & c \\ -c & d & a & -b \\ -d & -c & b & a \end{bmatrix} = \begin{bmatrix} \begin{bmatrix} a & b \\ -b & a \end{bmatrix} & \begin{bmatrix} c & d \\ -d & c \end{bmatrix} \\ \begin{bmatrix} -c & d \\ -d & -c \end{bmatrix} & \begin{bmatrix} a & -b \\ b & a \end{bmatrix} \end{bmatrix} = \begin{bmatrix} a+\hat{i}b & c+\hat{i}d \\ -c+\hat{i}d & a-\hat{i}b \end{bmatrix} \sim SU(2) \quad (4.10)$$

Indeed, Weinberg[16] does exactly this and calls the matrix the $SU(2)$ matrix.

We gain nothing from using complex numbers in matrices except that we save a little paper. However, we introduce a level of obscuration which hangs around us like a smog choking our understanding; for example, the reader might think the below matrix, (4.11), is a symmetric matrix:

$$\text{Symmetric matrix} \quad ??? \quad \begin{bmatrix} 0 & \hat{i} \\ \hat{i} & 0 \end{bmatrix} \quad (4.11)$$

This, (4.11), is an anti-symmetric matrix:

$$\begin{bmatrix} 0 & \hat{i} \\ \hat{i} & 0 \end{bmatrix} = \begin{bmatrix} 0 & 0 & 0 & 1 \\ 0 & 0 & -1 & 0 \\ 0 & 1 & 0 & 0 \\ -1 & 0 & 0 & 0 \end{bmatrix} \quad (4.12)$$

Anti-symmetric matrices, written using only real numbers, always have imaginary eigenvalues. Symmetric matrices, written using only real numbers, always have real eigenvalues. When the physicist says he wants hermitian matrices, what he really wants is matrices with real eigenvalues – symmetric matrices (written with only real numbers). We are able to throw hermitian matrices into the bin if we use only real numbers in our matrices and use symmetric matrices instead of hermitian matrices. That's one bit of the smog lifted. Beware though, the symmetry is in the notation and not in the matrix. I could write a 2×2 matrix as four numbers scattered randomly upon a bit of paper. The 'symmetric' matrix has no symmetry, but it does have real eigenvalues. It is the notation that has the symmetry. It is the matrix that has the real eigenvalues. We would be better talking of real eigenvalue matrices rather than symmetric matrices or hermitian matrices.

Summary:

We have seen both the quaternions and the complex numbers arise as finite group spaces. This might have shocked the reader who thought that the complex numbers were an algebraic extension of the real numbers based upon a monic minimum polynomial[17].

Did you see how simple it all is? What a spiffingly fine adventure we are having, wot! wot!

---

[16] Steven Weinberg : The Quantum Theory of Fields.
[17] If you do not know what an algebraic extension based on a monic minimum polynomial is, you are lucky. You do not need to know anything of this to understand this book.

Chapter 5

Division Algebras

We have seen that two the division algebras which are the 2-dimensional Euclidean complex numbers, $\mathbb{C}$, and the 4-dimensional quaternions, $\mathbb{H}$, have emerged from representations of the finite groups. To this we can add the 1-dimensional real numbers, $\mathbb{R}$, which, trivially, emerge from the finite group of order one, C_1, which is just the identity:

$$C_1 = [1] \qquad \exp([a]) = e^{[a]}e^{[0]} \qquad e^{[0]} = [1] \qquad \det([a]) = a \qquad (5.1)$$

We see that the 1-dimensional trigonometric function is just the number one and that a 1-dimensional angle has zero variables and that the 1-dimensional distance function is just the variable.

Great, all three zero characteristic[18] division algebras are finite group spaces.

It seems unfair. Surely, all finite group spaces ought to be division algebras. Surely the 2-dimensional space-time which we observe to be part of physical reality ought to be a division algebra just as the 2-dimensional Euclidean space which we also observe to be part of physical reality is a division algebra.

The algebraic nature of finite group spaces:

Frobenius and Dedekind progressed as far as the algebraic matrix form which they called a representation space or a representation module, but, as far as I am aware, they did not take the exponential of the algebraic matrix form. Let us ask whether or not the algebraic matrix form of the order two cyclic finite group is a division algebra. We begin by writing this algebraic matrix form slightly differently:

$$C_2^{\text{Algebraic matrix form}} = \begin{bmatrix} t & z \\ z & t \end{bmatrix} = \begin{bmatrix} 1 & 0 \\ 0 & 1 \end{bmatrix} t + \begin{bmatrix} 0 & 1 \\ 1 & 0 \end{bmatrix} z = t + z\sqrt{+1} = \mathbb{S}$$

$$(5.2)$$

$$C_4^{\text{Algebraic matrix form}} = \begin{bmatrix} x & y \\ -y & x \end{bmatrix} = \begin{bmatrix} 1 & 0 \\ 0 & 1 \end{bmatrix} x + \begin{bmatrix} 0 & 1 \\ -1 & 0 \end{bmatrix} y = x + y\sqrt{-1} = \mathbb{C}$$

We have included a similar rewrite of the algebraic matrix form which is the 2-dimensional Euclidean complex numbers for comparison. We see that the algebraic matrix form of the 2-dimensional representation of the order two finite cyclic group C_2 has a square root of plus unity whilst the familiar complex numbers have a square root of minus unity.

---

[18] We have no interest in modular division algebras.

If we try to form a division algebra using a square root of plus unity, we find that we have zero divisors:

$$\left(1+\sqrt{+1}\right)\left(1-\sqrt{+1}\right)=1-1=0 \tag{5.3}$$

In (5.3), we have two non-zero elements of the proposed division algebra which multiply together to form zero; naughty, naughty. This will be the case for all proposed division algebras which contain a square root of plus unity. Can we avoid this problem? Perhaps we could limit the values of the imaginary variable in some way such as[19]:

$$\left(t+z\sqrt{+1}\right)\left(t-z\sqrt{+1}\right)=t^2-z^2>0 \quad : \quad z<t \;\; \forall \;\; \{t,z\} \tag{5.4}$$

Perhaps remarkably[20], taking the exponential does exactly this:

$$\exp\left(\begin{bmatrix} t & z \\ z & t \end{bmatrix}\right)=\begin{bmatrix} r & 0 \\ 0 & r \end{bmatrix}\begin{bmatrix} \cosh z & 0 \\ 0 & \cosh z \end{bmatrix}+\begin{bmatrix} r & 0 \\ 0 & r \end{bmatrix}\begin{bmatrix} 0 & \sinh z \\ \sinh z & 0 \end{bmatrix} \tag{5.5}$$

$$\cosh z > \sinh z \quad \forall \quad z$$

By taking the exponential, we solve the zero divisors problem. The way forward is obvious. We take our finite group space to be the polar form of the algebraic matrix form. This is the bit that Frobenius and Dedekind missed.

We have the hyperbolic complex numbers:

$$\mathbb{S}=\begin{bmatrix} r & 0 \\ 0 & r \end{bmatrix}\begin{bmatrix} \cosh\chi & \sinh\chi \\ \sinh\chi & \cosh\chi \end{bmatrix} \qquad \mathbb{C}=\begin{bmatrix} r & 0 \\ 0 & r \end{bmatrix}\begin{bmatrix} \cos\theta & \sin\theta \\ -\sin\theta & \cos\theta \end{bmatrix} \tag{5.6}$$

In (5.6), we have included the Euclidean complex numbers for comparison.

Unfortunately, the polar form of the hyperbolic complex numbers, $\mathbb{S}$, does not have additive inverses on the real axis. We have $e^t>0 \;\forall\; t$. Nor do we have an additive identity (zero) on the real axis. The fourteen axioms of an algebraic field include an axiom requiring additive inverses on the real axis and an axiom requiring an additive identity on the real axis. But 2-dimensional space-time really exists. What has gone wrong?

The axiomatic system:
Traditionally, mathematics is an axiomatic system. We begin with a set of axioms written in stone, and we make deductions from those axioms. The arbiter of truth within axiomatic mathematics is the axioms. Where do the axioms come from? Human beings invent them.

Hang on, how do we know that the set of axioms invented by human beings are the correct axioms? We do not.

---

[19] There is an associated problem. If the two variables in the C_2 algebraic matrix form are equal, then we have a singular matrix – no inverse.
[20] Your author is unsure whether or not this is remarkable, but it might be remarkable; hence this remark.

Historically, the set of algebraic field axioms was copied from the real numbers, $\mathbb{R}$, and the complex numbers, $\mathbb{C}$. If our axiom writing forebears had known about the hyperbolic complex numbers, $\mathbb{S}$, they might have chosen a different set of axioms which included the hyperbolic complex numbers within them as an algebraic field. That's what went wrong.

The reader might now be expecting that we are about to proclaim every finite group space to be an algebraic field under a justifiable rewrite of the algebraic field axioms. We could do this, but we will take a more drastic route. We reject the axiomatic system completely. An axiomatic system is a religion. Within Islam, the Koran is the set of axioms which is the arbiter of truth. Christianity has the Bible. Mathematicians have the thirty-four axioms of modern mathematics as their arbiter of truth. We are atheists.

A complete rewrite of mathematics:

While we are here, we might as well rewrite the whole of mathematics. We do it like this. We declare that the number one exists. Philosophers might take the view that the number one must exist for, if it did not exist, then there would be zero number ones and zero is a number; it is one number. We do not wish to get 'bogged down' in philosophical considerations.

We declare that the number one exists, and we declare that mathematics is all that follows from the existence of the number one. We take anything that does not follow from the existence of the number one, and we throw it in the bin.

This is not an idea of your author. A hundred years ago, Bertrand Russell (1872-1970) and Alfred North Whitehead (1861-1947) began constructing mathematics in this way. Bertrand Russell took fifteen years to show that, if the number one exists, then the real numbers exist. That's the hard part done.

Once we have the real numbers, we have distinct mathematical objects and we have the concept of order. We can order the distinct objects in different ways. Thus, we have permutations. Thus, we have the finite groups. Combine together the finite groups and the real numbers, and we have the finite group spaces.

That's it. The hyperbolic complex numbers, and all the other finite group spaces, in their polar form, are what we declare to be division algebras. Most of the finite group spaces lack additive inverses on the real axis and lack an additive identity of the real axis, but this is acceptable. Actually, it is comforting to know there is no zero on the time axis of 2-dimensional space-time[21] because it implies that there was no zero time point.

There are three 'special' finite group spaces which have additive inverses on the real axis and an additive identity on the real axis; these are the two types of quaternions and the 2-dimensional Euclidean complex numbers. These are included in our list of division algebras, of course.

One of the remarkable things about constructing mathematics from no more than the existence of the number one is that a great deal of traditional mathematics, number theory, finite group theory, the complex numbers, etc., can be built upon no more than the existence of the number one.

---

[21] This is an emotional outburst and not a mathematical fact.

Another remarkable thing about constructing mathematics from no more than the existence of the number one is that a great deal of traditional mathematics is dumped in the skip. We are rid of Lie algebras, Riemannian geometry except for our 4-dimensional space-time, Hilbert spaces and vector spaces in general but excluding our 4-dimensional space-time.

We take the view that a vector space is an ordered n-tuplet of numbers, $\mathbb{R}^n$ or $\mathbb{C}^n$, and that the finite group spaces are comprised of a single real axis and a number of imaginary axes – we do not consider the finite group spaces to be vector spaces[22] formed of ordered n-tuplets of numbers.

The algebraic nature of the finite group spaces:
Fully aware that the finite group spaces violate a couple of the traditional axioms that traditionally define a division algebra, we declare the finite group spaces to be division algebras. When we use the term division algebra in this book, we will mean finite group space. Two phrases for the same thing.

No more algebraic extensions:
Traditionally, the 2-dimensional complex numbers are seen as an algebraic extension of the real numbers. We have a monic[23] polynomial:

$$x^2 + 1 = 0$$
$$\left(x + \hat{i}\right)\left(x - \hat{i}\right) = 0$$

(5.7)

This polynomial, (5.7), will not factorise into linear factors using only real numbers; we say it is a minimal polynomial because it has this property. To factorise this polynomial, we have to introduce the complex numbers, and so it is said that the complex numbers are an algebraic extension of the real numbers based upon this monic minimal polynomial, (5.7).

This view is not incorrect, but it is 'barking up the wrong tree'. In fact, it's running around in the wrong forest. We reject the algebraic extension view of the origin of the complex numbers, and we replace it with the finite group space view of the origin of the complex numbers.

Summary:
We have declared every finite group space to be a division algebra. We have firmly rooted division algebras within the finite groups. We have rejected the concepts of algebraic extensions as being 'running around in the wrong forest'. A finite group space is a division algebra.

There are problems associated with having adventures; you can easily get into trouble.

---

[22] This is clearly different from the accepted mantra.
[23] Monic means that the coefficient of the highest power of the variable in the polynomial is unity.

Chapter 6

The Algebraic Operations

To differentiate non-commutatively, we need some form of non-commutative multiplicative algebraic operation which will allow us to form the differential. In this chapter, we find that non-commutative multiplicative algebraic operation.

The standard mantra:

Abstract algebraists tell us that there are two algebraic operations called addition and multiplication. When we look at the real numbers, we might doubt this for is not multiplication within the real numbers nothing more than successive addition within the real numbers? However, when we look at the 2-dimensional complex numbers, $\mathbb{C}$, we certainly have two distinct algebraic operations; within the complex numbers, multiplication is not successive addition.

Great, it seems like the abstract algebraists understand these things correctly, but hang on; how do we know there are not three algebraic operations? What is an algebraic operation? Let us look at the 2-dimensional Euclidean complex numbers from a finite group space point of view. We have:

$$\mathbb{C} = \left\{ \begin{array}{cccc} \begin{bmatrix} 1 & 0 \\ 0 & 1 \end{bmatrix} & \begin{bmatrix} -1 & 0 \\ 0 & -1 \end{bmatrix} & \begin{bmatrix} 0 & 1 \\ -1 & 0 \end{bmatrix} & \begin{bmatrix} 0 & -1 \\ 1 & 0 \end{bmatrix} \\ \downarrow & \downarrow & \downarrow & \downarrow \\ 1 & -1 & \hat{i} & -\hat{i} \end{array} \right\} \tag{6.1}$$

Each of the parts of the complex number is really a permutation. This is how we formed the complex numbers. We took four permutations, and we multiplied each permutation by a real variable. The adding of these four permutations was a neat bit of notation, but, essentially, a complex number is a set of different amounts of four permutations.

Hm!, that is quite profound. Perhaps you should have a coffee break while you think about it.

What can we do with amounts of permutations? Well, we can add two permutations of the same kind; we call this addition:

$$a \begin{bmatrix} 1 & 0 \\ 0 & 1 \end{bmatrix} + c \begin{bmatrix} 1 & 0 \\ 0 & 1 \end{bmatrix} = (a+c) \begin{bmatrix} 1 & 0 \\ 0 & 1 \end{bmatrix}$$

$$\left(a \begin{bmatrix} 1 & 0 \\ 0 & 1 \end{bmatrix} + b \begin{bmatrix} 0 & 1 \\ -1 & 0 \end{bmatrix} \right) + \left(c \begin{bmatrix} 1 & 0 \\ 0 & 1 \end{bmatrix} + d \begin{bmatrix} 0 & 1 \\ -1 & 0 \end{bmatrix} \right) = \left((a+c) \begin{bmatrix} 1 & 0 \\ 0 & 1 \end{bmatrix} + (b+d) \begin{bmatrix} 0 & 1 \\ -1 & 0 \end{bmatrix} \right) \tag{6.2}$$

$$\left(a + \hat{i}b \right) + \left(c + \hat{i}d \right) = (a+c) + \hat{i}(b+d)$$

This is the algebraic operation of addition – we just add amounts of the same permutation together.

We can also combine permutations together sequentially. We begin with, say, 4 coloured balls in a given order, and we swap a few balls to form another order; we then form a third order by swapping some balls in the second order. This is the sequential combination of permutations. We call this multiplication:

$$\left(a\begin{bmatrix} 1 & 0 \\ 0 & 1 \end{bmatrix} + b\begin{bmatrix} 0 & 1 \\ -1 & 0 \end{bmatrix} \right) \left(c\begin{bmatrix} 1 & 0 \\ 0 & 1 \end{bmatrix} + d\begin{bmatrix} 0 & 1 \\ -1 & 0 \end{bmatrix} \right)$$

$$= ac\begin{bmatrix} 1 & 0 \\ 0 & 1 \end{bmatrix}\begin{bmatrix} 1 & 0 \\ 0 & 1 \end{bmatrix} + ad\begin{bmatrix} 1 & 0 \\ 0 & 1 \end{bmatrix}\begin{bmatrix} 0 & 1 \\ -1 & 0 \end{bmatrix} + bc\begin{bmatrix} 0 & 1 \\ -1 & 0 \end{bmatrix}\begin{bmatrix} 1 & 0 \\ 0 & 1 \end{bmatrix} + bd\begin{bmatrix} 0 & 1 \\ -1 & 0 \end{bmatrix}\begin{bmatrix} 0 & 1 \\ -1 & 0 \end{bmatrix}$$

(6.3)

$$= (ac - bd)\begin{bmatrix} 1 & 0 \\ 0 & 1 \end{bmatrix} + (ad + bc)\begin{bmatrix} 0 & 1 \\ -1 & 0 \end{bmatrix}$$

$$= (a + \hat{i}b)(c + \hat{i}d) = (ac - bd) + \hat{i}(ad + bc)$$

Algebraic multiplication is the sequential combination of permutations. Perhaps we need to emphasize what we have just found:

Algebraic addition is the adding of amounts of the same permutation.

(6.4)

Algebraic multiplication is the sequential combining together of two permutations.

Because the sequential combination of permutations can be non-commutative, multiplication can be non-commutative, as it is in the quaternions. Of course, addition is never non-commutative.

There we have it. An algebraic operation is two permutations in and one permutation out.

Why there are only two algebraic operations:
There are two algebraic operations because there are two ways of combining together permutations. Great, now we understand the nature of algebraic operations. Obviously, there are only two ways to combine together two permutations, and so, obviously, there are only two algebraic operations.

Let us just look at the real numbers:

Addition $\qquad\qquad a[1] + b[1] = (a + b)[1]$

(6.5)

Multiplication $\qquad\qquad a[1]b[1] = ab[1]$

We see that because the real numbers are based upon the single identity permutation, it seems like there is only one algebraic operation within the real numbers even though there are really two algebraic operations within the real numbers.

A third algebraic operation:

It is obvious that there can be only two algebraic operations because there are only two ways to put two permutations in and get one permutation out. Obvious it might be; true it is not. Let us look at the quaternions. Remember, the quaternions are non-commutative. In traditional notation (watch the signs):

$$\left(0a + b\hat{i} + cj + 0d\right)\left(0e + f\hat{i} + gj + 0h\right) = \left(-bf - cg\right) + 0\hat{i} + 0j + \left(bg - cf\right)k$$

(6.6)

$$\left(0e + f\hat{i} + gj + 0h\right)\left(0a + b\hat{i} + cj + 0d\right) = \left(-bf - cg\right) + 0\hat{i} + 0j - \left(bg - cf\right)k$$

In matrices, using the left-chiral quaternions, this is (signs):

$$\begin{bmatrix} 0 & b & c & 0 \\ -b & 0 & 0 & c \\ -c & 0 & 0 & -b \\ 0 & -c & b & 0 \end{bmatrix} \begin{bmatrix} 0 & f & g & 0 \\ -f & 0 & 0 & g \\ -g & 0 & 0 & -f \\ 0 & -g & f & 0 \end{bmatrix} = \begin{bmatrix} -bf - cg & 0 & 0 & bg - cf \\ 0 & -bf - cg & bg - cf & 0 \\ 0 & bg - cf & -bf - cg & 0 \\ bg - cf & 0 & 0 & -bf - cg \end{bmatrix}$$

(6.7)

$$\begin{bmatrix} 0 & f & g & 0 \\ -f & 0 & 0 & g \\ -g & 0 & 0 & -f \\ 0 & -g & f & 0 \end{bmatrix} \begin{bmatrix} 0 & b & c & 0 \\ -b & 0 & 0 & c \\ -c & 0 & 0 & -b \\ 0 & -c & b & 0 \end{bmatrix} = \begin{bmatrix} -bf - cg & 0 & 0 & -bg + cf \\ 0 & -bf - cg & -bg + cf & 0 \\ 0 & -bg + cf & -bf - cg & 0 \\ -bg + cf & 0 & 0 & -bf - cg \end{bmatrix}$$

What about the operation:

$$\left[\mathbb{H}_1, \quad \mathbb{H}_2\right] = \frac{1}{2}\left(\mathbb{H}_1\mathbb{H}_2 - \mathbb{H}_2\mathbb{H}_1\right)$$

(6.8)

This, (6.8), is called the commutator. The commutator is found within Lie algebras, but we've dumped Lie algebra in the skip. However, we might think the commutator could be an algebraic operation because of the established mathematical identity[24]:

$$e^A e^B = e^{A + B + \frac{1}{2}[A,B] + \ldots + \{A,B\} + \ldots}$$

(6.9)

$\{A, B\}$ are matrices which are not neccessarily commutative.

---

[24] This is the Baker-Campbell-Hausdorff theorem.

The product of the exponentials of two matrices is a function of the commutator. This looks very algebraic.

Using the commutator, we get:

$$\left[\left(0a+b\hat{i}+cj+0d\right),\left(0e+f\hat{i}+gj+0h\right)\right]=\frac{1}{2}\left(\left(-bf+bgk-cfk-cg\right)-\left(-bf-bgk+cfk-cg\right)\right)$$

$$=(bg-cf)k$$

(6.10)

In matrix form, this commutator operation is:

The Commutator of two matrices

(6.11)

$$\left[\begin{bmatrix} 0 & b & c & 0 \\ -b & 0 & 0 & c \\ -c & 0 & 0 & -b \\ 0 & -c & b & 0 \end{bmatrix}, \begin{bmatrix} 0 & f & g & 0 \\ -f & 0 & 0 & g \\ -g & 0 & 0 & -f \\ 0 & -g & f & 0 \end{bmatrix}\right] = \begin{bmatrix} 0 & 0 & 0 & bg-cf \\ 0 & 0 & bg-cf & 0 \\ 0 & bg-cf & 0 & 0 \\ bg-cf & 0 & 0 & 0 \end{bmatrix}$$

We have here two permutations in and one permutation out.

But two permutations in and one permutation out is an algebraic operation!

Within the quaternions, we have a third algebraic operation. Within the quaternions, the commutator is a third algebraic operation.

Emphasis:
An algebraic operation is two permutations in and one permutation out. Within the quaternions, we have the commutator operation which is of the form two permutations in and one permutation out. We have a third algebraic operation within the quaternions.

The importance of the commutator and the anti-commutator:
Along with the commutator, we have the anti-commutator:

$$\text{Anti-commutator} \qquad \{A, \ B\} = \frac{1}{2}(AB+BA)$$

(6.12)

$$\text{Commutator} \qquad [A, \ B] = \frac{1}{2}(AB-BA)$$

$A \& B$ are both square matrices.

These are *bona fide* algebraic operations – two permutations in, and one permutation out.

The importance of the commutator and the anti-commutator being algebraic operations is that there is a differential associated with them. Because the commutator and the anti-commutator are *bona fide* algebraic operations, we can form a non-commutative derivative based on the commutator operation. This means we can derive divergences and curls in finite group spaces like the quaternions. We will devote a later chapter to this. Divergences and curls mean fields, and fields mean that we have a potential. There is such a thing as a quaternion potential.

The commutator in commutative finite group spaces:
The commutator is something other than zero in the quaternions because the quaternions are non-commutative. Within a commutative finite group space like the complex numbers, $\mathbb{C}$, the anti-commutator is just normal multiplication and the commutator is zero:

$$\left\{ \begin{bmatrix} a & b \\ -b & a \end{bmatrix}, \begin{bmatrix} c & d \\ -d & c \end{bmatrix} \right\} = \frac{1}{2}\left(\begin{bmatrix} ac-bd & ad+bc \\ -(ad+bc) & ac-bd \end{bmatrix} + \begin{bmatrix} ac-bd & ad+bc \\ -(ad+bc) & ac-bd \end{bmatrix} \right)$$

(6.13)

$$= \begin{bmatrix} ac-bd & ad+bc \\ -(ad+bc) & ac-bd \end{bmatrix}$$

$$\left[\begin{bmatrix} a & b \\ -b & a \end{bmatrix}, \begin{bmatrix} c & d \\ -d & c \end{bmatrix} \right] = \frac{1}{2}\left(\begin{bmatrix} ac-bd & ad+bc \\ -(ad+bc) & ac-bd \end{bmatrix} - \begin{bmatrix} ac-bd & ad+bc \\ -(ad+bc) & ac-bd \end{bmatrix} \right)$$

(6.14)

$$= 0$$

We see that the commutator operation sub-cludes the multiplication operation. We might say that the multiplication operation is not an independent operation but is a shortened version of the commutator operation. We can use the commutator operation within commutative division algebras and dispense entirely with the normal multiplication operation. This has implications for forming the derivative in division algebras which are only partly non-commutative[25]; it means that we can form the derivative in such division algebras.

The absence of the commutator in the 8-dimensional quaternion finite group space:
Remember the commutator, and the anti-commutator, exist in the 4-dimensional representation of the order eight quaternion group; this is a 4-dimensional finite group space.

How about the commutator in the 8-dimensional representation of the quaternion group? This is an 8-dimensional finite group space. The commutator and the anti-commutator do not exist in the 8-dimensional representation of the quaternion group. The 4-dimensional left-chiral quaternion representation is:

---

[25] The non-commutative 8-dimensional $C_2 \times C_2 \times C_2$ algebras have a commutative real variable and one wholly commutative imaginary variable and six non-commutative variables.

$$Q_8^{LX} = \left\{ \begin{bmatrix} 1 & 0 & 0 & 0 \\ 0 & 1 & 0 & 0 \\ 0 & 0 & 1 & 0 \\ 0 & 0 & 0 & 1 \end{bmatrix}, \begin{bmatrix} 0 & 1 & 0 & 0 \\ -1 & 0 & 0 & 0 \\ 0 & 0 & 0 & -1 \\ 0 & 0 & 1 & 0 \end{bmatrix}, \begin{bmatrix} 0 & 0 & 1 & 0 \\ 0 & 0 & 0 & 1 \\ -1 & 0 & 0 & 0 \\ 0 & -1 & 0 & 0 \end{bmatrix}, \begin{bmatrix} 0 & 0 & 0 & 1 \\ 0 & 0 & -1 & 0 \\ 0 & 1 & 0 & 0 \\ -1 & 0 & 0 & 0 \end{bmatrix} \right.$$
$$\left. \begin{bmatrix} -1 & 0 & 0 & 0 \\ 0 & -1 & 0 & 0 \\ 0 & 0 & -1 & 0 \\ 0 & 0 & 0 & -1 \end{bmatrix}, \begin{bmatrix} 0 & -1 & 0 & 0 \\ 1 & 0 & 0 & 0 \\ 0 & 0 & 0 & 1 \\ 0 & 0 & -1 & 0 \end{bmatrix}, \begin{bmatrix} 0 & 0 & -1 & 0 \\ 0 & 0 & 0 & -1 \\ 1 & 0 & 0 & 0 \\ 0 & 1 & 0 & 0 \end{bmatrix}, \begin{bmatrix} 0 & 0 & 0 & -1 \\ 0 & 0 & 1 & 0 \\ 0 & -1 & 0 & 0 \\ 1 & 0 & 0 & 0 \end{bmatrix} \right\} \quad (6.15)$$

Notice how the permutations come in pairs which are negatives of each other. This is an essential property of a representation for the commutator to exist. Just think about it for a while. Within the quaternions, we have $\hat{i}j = k$ but we also have $\hat{ji} = -k$. The non-commutativity is in the sign. We need a $-k$ to match every k.

Within the commutator operation, we add/subtract two different permutations. We cannot add different permutations together unless one is the negative of the other. I really do want you to think about this.

The 8-dimensional representation of the order eight quaternion finite group is eight 8×8 matrices with only zeros and plus-ones. There are no minus-ones in the 8-dimensional representation of the quaternion group. No permutation in the 8-dimensional representation of the quaternion group is a negative of another permutation.

The commutator is not an algebraic operation within the 8-dimensional representation of the order eight quaternion finite group, but the commutator is an algebraic operation in the 4-dimensional representation of the order eight quaternion group.

The commutator is an algebraic operation in only some representations of a finite group. Those representations must be of a dimension, matrix size, equal to half the order of the finite group represented. Those representations must have the 'negativity property' like the 4-dimensional representation of the order eight quaternion group.

This will mean that we can form the non-commutative differential in only some representations of some finite groups. This will mean we cannot form the curl of the finite group in most non-commutative finite group spaces or in most representations of most non-commutative finite groups.

The absence of the commutator in non-commutative groups in general:
Consider the representation of any finite group which is of the same dimension as the order of the finite group. That representation is formed of only zeros and plus-ones. There can be no negative permutations in such a representation, and so there is no commutator operation in such a finite group space.

In representations of any non-commutative finite group which is of the same dimension as the order of the finite group, the non-commutativity is expressed as different permutations not as the sign of the permutation.

In the next chapter, we will make clearer which representations hold the commutator operation.

Summary:

We have seen that the two algebraic operations of addition and multiplication are really just two ways of combining together two permutations to make a third permutation.

We have shown that, in some 'special' representations of non-commutative finite groups, that is in some 'special' finite group spaces, that is in some 'special' division algebras, there is a third algebraic operation which we call the commutator and the anti-commutator.

We have shown that the commutator, which is extensively used in particle physics, is actually a proper *bona fide* algebraic operation within some 'special' division algebras. These 'special' division algebras do not include Lie algebras which we have dumped in the skip.

A preview:

We will soon see that every commutative finite group space has a curl. We will see that the number of terms in the curl of a commutative finite group space is equal to the dimension of the finite group space. The two 2-dimensional finite group spaces we have met above, the 2-dimensional Euclidean space and the 2-dimensional space-time, are the only commutative finite group spaces that have curls with two and only two terms.

We will soon see that most non-commutative finite group spaces do not have a divergence or a curl. We will see that, because the commutator operation exists in some 'special' non-commutative finite group spaces, those 'special' non-commutative finite group spaces have a curl. In some instances, these curls can have two, and only two, terms. That means they can be manifest in our 4-dimensional space-time. A curl is associated with a force. A divergence is associated with a charge. We are starting to see why we have the forces we do have in our universe.

Chapter 7

From Commutative Representations to the Commutator

We are particularly interested in finite group spaces which have the commutator as an algebraic operation. These finite group spaces will necessarily be representations of dimension, matrix size, half the order of the finite group which they represent and they will necessarily be square matrices formed of zeros, plus-ones, and minus-ones.

The essence of the commutator is that the non-commutativity of the representation is in the sign of the product of two elements of the finite group representation. If the non-commutativity is in the variable then the commutator does not exist within this finite group representation. :

$$ab = c \qquad\qquad ba = d$$

$$(7.1)$$

Non-commutativity is in the variable

We need:

$$ab = c \qquad\qquad ba = -c$$

$$(7.2)$$

Non-commutativity is in the sign

The order of the elements of the group:
Consider a representation of a non-commutative finite group which is of half the order of the non-commutative finite group and which has relations that support the commutator like:

$$BC = D \quad CB = -D \quad CD = B \quad DC = -B \quad DB = C \quad BD = -C \qquad (7.3)$$

Consider:

$$BBC = BD = -C \quad \Rightarrow \quad BB = -I \qquad (7.4)$$

We see that the element of the group B is the square root of minus the identity. This means that the element B must be of order four within the group. The same applies to all other elements in the above list, (7.3).

Now let us change the signs a little:

$$BC = -D \quad CB = D \quad CD = B \quad DC = -B \quad DB = C \quad BD = -C \qquad (7.5)$$

This rearrangement can still support the commutator. Consider:

$$BBC = -BD = C \quad \Rightarrow \quad BB = I \qquad (7.6)$$

We see that the element of the group B is now the square root of plus the identity. This means that the element B must be of order two within the group, (7.5).

The conclusion is that the commutator can consistently exist within a finite group space of dimension N, division algebra of dimension N, only if that finite group space is a representation of a finite group of order $2N$ such that every element in that finite group is of order one (the single identity) or of order two or of order four. If we see a finite group which has elements of any order other than two or four (or one), then we know that finite group does not have a representation which will support the commutator as an algebraic operation.

The reader is urged to contemplate the previous few paragraphs. They mean there is a shortage of finite group spaces which support the commutator and thus a shortage of finite group spaces that might harbour curls and divergences that can be manifest in our 4-dimensional space-time.

The calculation:
How do we find these finite group representations which have the commutator as an algebraic operation?

We are not particularly interested in the 4-dimensional representations of the order four cyclic finite group C_4, but we will use it to illustrate the calculation of some 4-dimensional representations of order eight groups. We begin with the 4-dimensional representation of the order four cyclic finite group. Since this representation is formed from matrices of the same size as the order of the group, we will form it using only zeros and plus-ones. We have:

$$C_4 = \left\{ \begin{bmatrix} 1 & 0 & 0 & 0 \\ 0 & 1 & 0 & 0 \\ 0 & 0 & 1 & 0 \\ 0 & 0 & 0 & 1 \end{bmatrix}, \begin{bmatrix} 0 & 1 & 0 & 0 \\ 0 & 0 & 1 & 0 \\ 0 & 0 & 0 & 1 \\ 1 & 0 & 0 & 0 \end{bmatrix}, \begin{bmatrix} 0 & 0 & 1 & 0 \\ 0 & 0 & 0 & 1 \\ 1 & 0 & 0 & 0 \\ 0 & 1 & 0 & 0 \end{bmatrix}, \begin{bmatrix} 0 & 0 & 0 & 1 \\ 1 & 0 & 0 & 0 \\ 0 & 1 & 0 & 0 \\ 0 & 0 & 1 & 0 \end{bmatrix} \right\} \tag{7.7}$$

From this, we form the corresponding algebraic matrix form:

$$C_4 = \begin{bmatrix} a & b & c & d \\ d & a & b & c \\ c & d & a & b \\ b & c & d & a \end{bmatrix} \tag{7.8}$$

We will now attach a parameter, $P_{i,j}$, to each element of the matrix. For our present purposes[26], these parameters can take only the values of plus unity or minus unity, ± 1.

---

[26] Within the representation spaces, finite group spaces, that seem to form our universe, we allow these parameters to take any real value except zero provided they do not change sign. It seems that these parameters, in the appropriate finite group spaces, are the physical constants of the universe.

$$C_4 = \begin{bmatrix} P_{1,1}a & P_{1,2}b & P_{1,3}c & P_{1,4}d \\ P_{2,1}d & P_{2,2}a & P_{2,3}b & P_{2,4}c \\ P_{3,1}c & P_{3,2}d & P_{3,3}a & P_{3,4}b \\ P_{4,1}b & P_{4,2}c & P_{4,3}d & P_{4,4}a \end{bmatrix} \tag{7.9}$$

We are going to seek multiplicative closure of the form of the matrix (7.9).

Pretty soon, we realise that, for each of the four separate variables in (7.9), we could divide throughout the matrix by the appropriate parameter on the top row, and replace the fractions so formed by a new single parameter, and so we might as well set the parameters on the top row to plus unity[27].

We are going to form a representation of a finite group. All finite groups have an identity. This implies that the parameters on the leading diagonal must be equal and positive. We set these to plus unity[28]. We get:

$$C_4 = \begin{bmatrix} a & b & c & d \\ P_{2,1}d & a & P_{2,3}b & P_{2,4}c \\ P_{3,1}c & P_{3,2}d & a & P_{3,4}b \\ P_{4,1}b & P_{4,2}c & P_{4,3}d & a \end{bmatrix} \tag{7.10}$$

We now form a copy of this C_4 algebraic matrix form, (7.10), and we form the product:

$$\begin{bmatrix} a & b & c & d \\ P_{2,1}d & a & P_{2,3}b & P_{2,4}c \\ P_{3,1}c & P_{3,2}d & a & P_{3,4}b \\ P_{4,1}b & P_{4,2}c & P_{4,3}d & a \end{bmatrix} \begin{bmatrix} e & f & g & h \\ P_{2,1}h & e & P_{2,3}f & P_{2,4}g \\ P_{3,1}g & P_{3,2}h & e & P_{3,4}f \\ P_{4,1}f & P_{4,2}g & P_{4,3}h & e \end{bmatrix} =$$

$$\begin{bmatrix} ae + P_{2,1}bh + P_{3,1}cg + P_{4,1}df & \sim & \sim & \sim \\ \sim & ae + P_{2,3}P_{3,2}bh + P_{2,4}P_{4,2}cg + P_{2,1}df & \sim & \sim \\ \sim & \sim & ae + P_{3,4}P_{4,3}bh + \sim + \sim & \sim \\ \sim & \sim & \sim & \sim \end{bmatrix}$$

$$(7.11)$$

We have squiggled out a few parts to fit the product on the page.

To have multiplicative closure of form, the elements on the leading diagonal must be equal. Thus, we can eliminate some parameters:

$$P_{3,2} = \frac{P_{2,1}}{P_{2,3}} \qquad\qquad P_{4,1} = P_{2,1} \qquad\qquad P_{4,2} = \frac{P_{3,1}}{P_{2,4}} \qquad\qquad P_{4,3} = \frac{P_{2,1}}{P_{3,4}} \tag{7.12}$$

---

[27] If we did not do this at the start, the calculation would force us to do this later. We simply avoid a laborious calculation.
[28] If we did not do this at the start, the calculation would force us to do this later. We simply avoid a laborious calculation.

We now have:

$$
\begin{bmatrix}
a & b & c & d \\
P_{2,1}d & a & P_{2,3}b & P_{2,4}c \\
P_{3,1}c & \dfrac{P_{2,1}}{P_{2,3}}d & a & P_{3,4}b \\
P_{2,1}b & \dfrac{P_{3,1}}{P_{2,4}}c & \dfrac{P_{2,1}}{P_{3,4}}d & a
\end{bmatrix}
\begin{bmatrix}
e & f & g & h \\
P_{2,1}h & e & P_{2,3}f & P_{2,4}g \\
P_{3,1}g & \dfrac{P_{2,1}}{P_{2,3}}h & e & P_{3,4}f \\
P_{2,1}f & \dfrac{P_{3,1}}{P_{2,4}}g & \dfrac{P_{2,1}}{P_{3,4}}h & e
\end{bmatrix}
\tag{7.13}
$$

Using comparison of the off-diagonal elements eventually leads to the multiplicatively closed algebraic matrix form:

$$
C_4 =
\begin{bmatrix}
a & b & c & d \\
P_{2,1}d & a & \dfrac{P_{2,1}P_{2,4}}{P_{3,1}}b & P_{2,4}c \\
P_{3,1}c & \dfrac{P_{3,1}}{P_{2,4}}d & a & P_{2,4}b \\
P_{2,1}b & \dfrac{P_{3,1}}{P_{2,4}}c & \dfrac{P_{2,1}}{P_{2,4}}d & a
\end{bmatrix}
\tag{7.14}
$$

Throughout the whole parameter elimination process, we were presented with only linear parameter elimination equations – never a quadratic parameter elimination equation. Each parameter elimination equation had only one solution, and so we arrived at only one multiplicatively closed algebraic matrix form, (7.14).

We now set the parameters, $P_{i,j} = \pm 1$. There are three parameters, and so there are $2^3 = 8$ possible permutations of the parameters as plus or minus one; this gives eight finite group spaces. Of these eight finite group spaces, the one with all parameters equal to plus unity is both a 4-dimensional representation of the order four cyclic finite group C_4 (all variables positive) and a representation of the order eight cyclic finite group C_8 (variables can take any value). The other seven finite group spaces are all 4-dimensional representations of the order eight cyclic finite group C_8. Note that there are sixteen more 4-dimensional representations of the order eight cyclic group because there are two more 4-dimensional representations of the order four cyclic group, (2.4) & (2.5).

Well, that's the idea. We are going to repeat this procedure with the only other order four finite group which is the Klein group $C_2 \times C_2$; this is a direct product group. Remember that there are only two order four finite groups.

Perhaps the most important finite group in the universe:
There is one and only one 4-dimensional representation of the order four direct product finite group $C_2 \times C_2$; this group is also known as the Klein group or *viergruppen*.

The single 4-dimensional representation of $C_2 \times C_2$ is:

$$C_2 \times C_2 = \left\{ \begin{bmatrix} 1 & 0 & 0 & 0 \\ 0 & 1 & 0 & 0 \\ 0 & 0 & 1 & 0 \\ 0 & 0 & 0 & 1 \end{bmatrix}, \begin{bmatrix} 0 & 1 & 0 & 0 \\ 1 & 0 & 0 & 0 \\ 0 & 0 & 0 & 1 \\ 0 & 0 & 1 & 0 \end{bmatrix}, \begin{bmatrix} 0 & 0 & 1 & 0 \\ 0 & 0 & 0 & 1 \\ 1 & 0 & 0 & 0 \\ 0 & 1 & 0 & 0 \end{bmatrix}, \begin{bmatrix} 0 & 0 & 0 & 1 \\ 0 & 0 & 1 & 0 \\ 0 & 1 & 0 & 0 \\ 1 & 0 & 0 & 0 \end{bmatrix} \right\} \tag{7.15}$$

The associated algebraic matrix form is:

$$C_2 \times C_2 = \begin{bmatrix} a & b & c & d \\ b & a & d & c \\ c & d & a & b \\ d & c & b & a \end{bmatrix} \tag{7.16}$$

We will do as we did above with the group C_4. Inserting the parameters gives:

$$C_2 \times C_2 = \begin{bmatrix} a & b & c & d \\ P_{2,1}b & a & P_{2,3}d & P_{2,4}c \\ P_{3,1}c & P_{3,2}d & a & P_{3,4}b \\ P_{4,1}d & P_{4,2}c & P_{4,3}b & a \end{bmatrix} \tag{7.17}$$

We find that we are able to eliminate five of the nine parameters using linear parameter elimination equations. This gives:

$$C_2 \times C_2 = \begin{bmatrix} a & b & c & d \\ P_{2,1}b & a & P_{2,3}d & \dfrac{P_{2,1}}{P_{2,3}}c \\ P_{3,1}c & \dfrac{P_{4,1}}{P_{2,3}}d & a & \dfrac{P_{2,3}P_{3,1}}{P_{4,1}}b \\ P_{4,1}d & \dfrac{P_{2,3}P_{3,1}}{P_{2,1}}c & \dfrac{P_{2,1}P_{4,1}}{P_{2,3}P_{3,1}}b & a \end{bmatrix} \tag{7.18}$$

We still have four free parameters. This, (7.18), is not a multiplicatively closed matrix form. We need to eliminate one more parameter to gain multiplicative closure[29]. How-so-ever we try, we discover that we are driven to a quadratic parameter elimination equation. That equation (for this particular choice of free parameters[30]) is:

$$P_{4,1}^{\,2} = \frac{P_{2,3}^{\,4} P_{3,1}^{\,2}}{P_{2,1}^{\,2}} \qquad P_{4,1} = +\frac{P_{2,3}^{\,2} P_{3,1}}{P_{2,1}} \qquad \text{or} \qquad P_{4,1} = -\frac{P_{2,3}^{\,2} P_{3,1}}{P_{2,1}} \tag{7.19}$$

---

[29] Within all finite group spaces, it seems that we get multiplicative closure of form for an order N group with (N - 1) free parameters. Why this is so is not clearly understood.
[30] We can choose to eliminate which-ever parameters we like to eliminate.

We have two solutions; we are driven to two algebraic matrix forms:

$$C_2 \times C_2^{\ Com} = \begin{bmatrix} a & b & c & d \\ P_{2,1}b & a & P_{2,3}d & \dfrac{P_{2,1}}{P_{2,3}}c \\ P_{3,1}c & \dfrac{P_{2,3}P_{3,1}}{P_{2,1}}d & a & \dfrac{P_{2,1}}{P_{2,3}}b \\ \dfrac{P_{2,3}^{\ 2}P_{3,1}}{P_{2,1}}d & \dfrac{P_{2,3}P_{3,1}}{P_{2,1}}c & P_{2,3}b & a \end{bmatrix} \tag{7.20}$$

And:

$$C_2 \times C_2^{\ Non-Com} = \begin{bmatrix} a & b & c & d \\ P_{2,1}b & a & P_{2,3}d & \dfrac{P_{2,1}}{P_{2,3}}c \\ P_{3,1}c & -\dfrac{P_{2,3}P_{3,1}}{P_{2,1}}d & a & -\dfrac{P_{2,1}}{P_{2,3}}b \\ -\dfrac{P_{2,3}^{\ 2}P_{3,1}}{P_{2,1}}d & \dfrac{P_{2,3}P_{3,1}}{P_{2,1}}c & -P_{2,3}b & a \end{bmatrix} \tag{7.21}$$

The difference between these two algebraic matrix forms, (7.20) & (7.21) is only four minus signs.

The first of these $C_2 \times C_2$ algebraic matrix forms is multiplicatively commutative. The eight permutations of $P_{i,j} = \pm 1$ give two representations of the order eight direct product finite group $C_2 \times C_2 \times C_2$, and six representations of the order eight direct product finite group $C_2 \times C_4$. These are both commutative finite groups.

The second of these $C_2 \times C_2$ algebraic matrix forms is multiplicatively closed but non-commutative. The non-commutativity is in the sign. The eight permutations of $P_{i,j} = \pm 1$ give six representations of the non-commutative order eight dihedral finite group D_4, and two representations of the order eight quaternion finite group Q_8. These are both non-commutative finite groups.

There are five, and only five, order eight finite groups, $\{C_8,\ C_2 \times C_2 \times C_2,\ C_2 \times C_4,\ D_4,\ Q_8\}$. The 4-dimensional representation of the order four cyclic group led to 4-dimensional representations of the order eight cyclic finite group C_8; the 4-dimensional representation of the order four direct product group $C_2 \times C_2$ has led to 4-dimensional representations of the remaining four order eight finite groups. We have all of the order eight finite groups as 4-dimensional representations. The reader might think this is a general phenomenon; it is not. The 3-dimensional representation of the order three cyclic group does not lead to the order six non-commutative symmetric group $S_3 \cong D_3$.

We now leave behind the $C_2 \times C_2$ commutative algebraic matrix form, (7.20). It is of little interest to us.

The commutator from commutative groups:

Because we began with a commutative group, $C_2 \times C_2$, and we came to representations of non-commutative groups, the commutator will exist in these non-commutative representations. The non-commutativity of these representations cannot be in the variable, because the basic algebraic matrix form is a commutative matrix based on a commutative finite group. The non-commutativity of these representations must be in the sign. The only difference between the commutative algebraic matrix form of the order four finite group $C_2 \times C_2$ and the non-commutative algebraic matrix form of the order four finite group $C_2 \times C_2$, (7.20) & (7.21) is four minus signs. The non-commutativity must be in the signs.

The difference in signs between the two algebraic matrix forms of the $C_2 \times C_2$ group derive from the quadratic parameter elimination equation.

This is how we find representations of non-commutative finite groups with the non-commutativity in the sign.

Summary:

We have seen how to begin with a commutative finite group of a given order, n, and calculate n-dimensional representations of finite groups of twice that given order. We have seen that, when a quadratic elimination equation necessarily arises, the basic algebraic matrix form separates into two different algebraic matrix forms of the same basic finite group. We have seen non-commutative groups arise from commutative groups. In some sense, some of the groups that exist at any given even order are, at least partially, determined by the nature of the groups of half that order. This is not clearly understood.

The important point is that we have found some commutator division algebras. Such algebras are quite rare. Most often, no quadratic parameter elimination equation arises within the parameter elimination process. The distribution of these commutator division algebras is not understood.

Addendum:

We derive the four finite group spaces which derive from the order three cyclic group C_3. These four finite group spaces are representations of the order six cyclic finite group C_6.

$$C_3 = \exp\left(\begin{bmatrix} a & b & c \\ c & a & b \\ b & c & a \end{bmatrix}\right)$$

$$dist^3 = a^3 + b^3 + c^3 - 3abc$$

(7.22)

This finite group space, (7.22), has two cube roots of plus unity

$$\exp\left(\begin{bmatrix} a & b & c \\ c & a & -b \\ b & -c & a \end{bmatrix}\right) \tag{7.23}$$

$$dist^3 = a^3 - b^3 - c^3 + 3abc$$

This finite group space, (7.23), has two cube roots of minus unity.

The remaining two representations of C_6 are:

$$\exp\left(\begin{bmatrix} a & b & c \\ -c & a & b \\ -b & -c & a \end{bmatrix}\right) \qquad \exp\left(\begin{bmatrix} a & b & c \\ -c & a & -b \\ -b & c & a \end{bmatrix}\right) \tag{7.24}$$

$$dist^3 = a^3 + b^3 - c^3 + 3abc \qquad dist^3 = a^3 - b^3 + c^3 + 3abc$$

These finite group spaces, (7.24), each have one cube root of plus unity and one cube root of minus unity. They are algebraically isomorphic.

Chapter 8

The 4-dim Reps. of the Order 8 Non-commutative Groups

We have the non-commutative algebraic matrix form which we derived from the order four finite group $C_2 \times C_2$:

$$C_2 \times C_2^{Non-Com} = \begin{bmatrix} a & b & c & d \\ P_{2,1}b & a & P_{2,3}d & \dfrac{P_{2,1}}{P_{2,3}}c \\ P_{3,1}c & -\dfrac{P_{2,3}P_{3,1}}{P_{2,1}}d & a & -\dfrac{P_{2,1}}{P_{2,3}}b \\ -\dfrac{P_{2,3}^{\;2}P_{3,1}}{P_{2,1}}d & \dfrac{P_{2,3}P_{3,1}}{P_{2,1}}c & -P_{2,3}b & a \end{bmatrix} \tag{8.1}$$

Choosing $P_{2,1} = P_{3,1} = P_{2,3} = -1$ gives:

$$\mathbb{H}_{L\chi} = \begin{bmatrix} a & b & c & d \\ -b & a & -d & c \\ -c & d & a & -b \\ -d & -c & b & a \end{bmatrix} \tag{8.2}$$

Note that if we had chosen to eliminate different parameters, some other combination of parameter values would have led to the same matrix form.

This, (8.2), is the left-chiral quaternions. The commutation relations of the variables in this matrix are the standard traditional commutation relations of the quaternions. We might refer to these commutation relations as $SU(2)_{L\chi}$.

Choosing $P_{2,1} = P_{3,1} = -1$ & $P_{2,3} = +1$ gives:

$$\mathbb{H}_{R\chi} = \begin{bmatrix} a & b & c & d \\ -b & a & d & -c \\ -c & -d & a & b \\ -d & c & -b & a \end{bmatrix} \tag{8.3}$$

Again, if we had chosen to eliminate different parameters, some other combination of parameter values would have led to the same matrix form.

This, (8.3), is the right-chiral quaternions. The commutation relations of the variables in this matrix are the reverse of the standard traditional commutation relations of the quaternions. We might refer to these commutation relations as $SU(2)_{R\chi}$.

Remember that the chirality of an algebra is in the commutation relations of that algebra. Looking at the above quaternions, (8.2) & (8.3), we see that the chirality is held in the distribution of minus signs in the bottom right-hand 3×3 corner of the matrices. The signs of the variables in the bottom right-hand 3×3 corner of the left-chiral quaternion matrix are opposite to the signs of the variables in the bottom right-hand 3×3 corner of the right-chiral quaternion matrix.

The determinant of the left-chiral quaternions is the same as the determinant of the right-chiral quaternions; this is another way of saying that they are similar matrices.

Electrons and neutrinos:
We will return to these two quaternion matrices later when we derive the electron and the neutrino fields. For now, our interest in in the six 4-dimensional representations of the order eight dihedral group which are known as the six A_3 algebras.

The A₃ algebras:
We refer to the six 4-dimensional representation spaces, finite group spaces, of the order eight dihedral group as the six A_3 algebras. We have:

$$SSA_{L\chi} = \exp\left(\begin{bmatrix} t & x & y & z \\ x & t & -z & -y \\ y & z & t & x \\ -z & -y & x & t \end{bmatrix}\right) \qquad SSA_{R\chi} = \exp\left(\begin{bmatrix} t & x & y & z \\ x & t & z & y \\ y & -z & t & -x \\ -z & y & -x & t \end{bmatrix}\right) \quad (8.4)$$

$$SAS_{L\chi} = \exp\left(\begin{bmatrix} t & x & y & z \\ x & t & z & y \\ -y & z & t & -x \\ z & -y & -x & t \end{bmatrix}\right) \qquad SAS_{R\chi} = \exp\left(\begin{bmatrix} t & x & y & z \\ x & t & -z & -y \\ -y & -z & t & x \\ z & y & x & t \end{bmatrix}\right) \quad (8.5)$$

$$ASS_{L\chi} = \exp\left(\begin{bmatrix} t & x & y & z \\ -x & t & -z & y \\ y & -z & t & -x \\ z & y & x & t \end{bmatrix}\right) \qquad ASS_{R\chi} = \exp\left(\begin{bmatrix} t & x & y & z \\ -x & t & z & -y \\ y & z & t & x \\ z & -y & -x & t \end{bmatrix}\right) \quad (8.6)$$

The names reflect the alphabetic order of the variables and the symmetric distribution or anti-symmetric distribution of each variable in the matrix. We see that each A_3 algebra has a real variable, the leading diagonal, two symmetric imaginary variables, and one anti-symmetric imaginary variable.

To be rid of zero-divisors, we need to take the exponential of the basic matrix form; this is because of the presence of the symmetric variables. The finite group spaces are the exponentials, polar form, of the basic matrix forms.

The chirality of our 4-dimensional space-time:

The set of commutation relations of our 4-dimensional space-time are known as the Lorentz group signified by $SO(3,1)$.

We see that we have both left chiral forms of the A_3 algebras and right-chiral forms of the A_3 algebras. Remember that the chirality is in the commutation relations. Taken together, the set of three left-chiral A_3 algebras has the commutation relations that are the Lorentz group, $SO(3,1)_{L\chi}$. Taken together, the set of three right-chiral A_3 algebras has the reverse commutation relations of the commutation relations that are the Lorentz group, $SO(3,1)_{R\chi}$. We have a left-chiral set of $SO(3,1)_{L\chi}$ commutation relations, and we have a right-chiral set of $SO(3,1)_{R\chi}$ commutation relations. We might suspect that these A_3 algebras are something to do with our 4-dimensional space-time – they are.

Consider the left-chiral $SSA_{L\chi}$ algebra with some variables set to zero. (We use matrices with positive variables on the top row):

$$
\begin{bmatrix} 0 & x & 0 & 0 \\ x & 0 & 0 & 0 \\ 0 & 0 & 0 & x \\ 0 & 0 & x & 0 \end{bmatrix}
\begin{bmatrix} 0 & 0 & 0 & z \\ 0 & 0 & -z & 0 \\ 0 & z & 0 & 0 \\ -z & 0 & 0 & 0 \end{bmatrix}
=
\begin{bmatrix} 0 & 0 & -xz & 0 \\ 0 & 0 & 0 & xz \\ -xz & 0 & 0 & 0 \\ 0 & xz & 0 & 0 \end{bmatrix}
$$

$$(8.7)$$

$$xz = -y$$

Now consider the right-chiral $SSA_{R\chi}$ algebra with some variables set to zero:

$$
\begin{bmatrix} 0 & x & 0 & 0 \\ x & 0 & 0 & 0 \\ 0 & 0 & 0 & -x \\ 0 & 0 & -x & 0 \end{bmatrix}
\begin{bmatrix} 0 & 0 & 0 & z \\ 0 & 0 & z & 0 \\ 0 & -z & 0 & 0 \\ -z & 0 & 0 & 0 \end{bmatrix}
=
\begin{bmatrix} 0 & 0 & xz & 0 \\ 0 & 0 & 0 & xz \\ xz & 0 & 0 & 0 \\ 0 & xz & 0 & 0 \end{bmatrix}
$$

$$(8.8)$$

$$xz = +y$$

We see that the chirality of these finite group spaces, as with the quaternions, is encoded in the distribution of minus signs bottom right-hand 3×3 corner of the matrices.

A₃ Rotation matrices:

We have:

$$\exp\left(\begin{bmatrix} t & x & y & z \\ -x & t & -z & y \\ y & -z & t & -x \\ z & y & x & t \end{bmatrix}\right) = \begin{bmatrix} r & 0 & 0 & 0 \\ 0 & r & 0 & 0 \\ 0 & 0 & r & 0 \\ 0 & 0 & 0 & r \end{bmatrix} \begin{bmatrix} \cosh\lambda & \dfrac{x}{\lambda}\sinh\lambda & \dfrac{y}{\lambda}\sinh\lambda & \dfrac{z}{\lambda}\sinh\lambda \\ -\dfrac{x}{\lambda}\sinh\lambda & \cosh\lambda & -\dfrac{z}{\lambda}\sinh\lambda & \dfrac{y}{\lambda}\sinh\lambda \\ \dfrac{y}{\lambda}\sinh\lambda & -\dfrac{z}{\lambda}\sinh\lambda & \cosh\lambda & -\dfrac{x}{\lambda}\sinh\lambda \\ \dfrac{z}{\lambda}\sinh\lambda & \dfrac{y}{\lambda}\sinh\lambda & \dfrac{x}{\lambda}\sinh\lambda & \cosh\lambda \end{bmatrix}$$

$$\lambda = \sqrt{-x^2 + y^2 + z^2}$$

(8.9)

This is a A_3 rotation matrix. The rotation matrices of the other A_3 algebras are basically the same as this rotation matrix, (8.9).

A₃ distance functions:

Ignoring the exponential, and taking the determinants of the six A_3 algebraic matrix forms above, (8.4) & (8.5) & (8.6) gives:

$$dist^4 = \left(t^2 - x^2 - y^2 + z^2\right)^2 \qquad dist^4 = \left(t^2 - x^2 - y^2 + z^2\right)^2$$

$$dist^4 = \left(t^2 - x^2 + y^2 - z^2\right)^2 \qquad dist^4 = \left(t^2 - x^2 + y^2 - z^2\right)^2 \qquad (8.10)$$

$$dist^4 = \left(t^2 + x^2 - y^2 - z^2\right)^2 \qquad dist^4 = \left(t^2 + x^2 - y^2 - z^2\right)^2$$

Perhaps the reader might sense what is coming next.

Summary:

This has been a short chapter in which we have presented the quaternion algebras and the A_3 algebras.

Chapter 9

Our 4-dimensional Space-time

In our exciting adventure, we now digress a little to prepare for what comes later. We will derive our 4-dimensional space-time from the A_3 finite group spaces.

What is our 4-dimensional space-time?

Our 4-dimensional space-time has the commutation relations of the Lorentz group $SO(3,1)$; everyone agrees about that, but it is utter nonsense. The multiplication operation exists within only division algebras (finite group spaces). Our 4-dimensional space-time is not a finite group space. Our 4-dimensional space-time is just four copies of the real numbers, $\mathbb{R}^4$. If there is no multiplication operation in our 4-dimensional space-time, there cannot be any commutation relations within our 4-dimensional space-time. Just what is happening here?

There are no angles within four copies of the real numbers, yet we observe angles within our 4-dimensional space-time. Only finite group spaces have angles. Just what is happening here?

We habitually differentiate within our 4-dimensional space-time by treating each dimension as a separate copy of the real numbers. The multiplication operation exists within the real numbers, and so we can form the differential within the real numbers. The real numbers are multiplicatively commutative, and so we form four copies of the commutative differential within out 4-dimensional space-time.

We would like to do non-commutative differentiation within our 4-dimensional space-time, but our 4-dimensional space-time is not a finite group space. We can differentiate within only division algebras (finite group spaces) because we need the algebraic operation to form the derivative. We need to understand the structure of our 4-dimensional space-time.

The concept of super-position:

We are going to form a super-position space. We do this by:

1) Choose a finite group.
2) Of the types of finite group space which are associated with the particular finite group, choose one type of finite group space.
3) Collect all algebraically isomorphic copies of that type finite group space.
4) Add the distance functions of all the copies of that finite group space.

We choose the A_3 finite group spaces. These are the six 4-dimensional representations of the non-commutative order eight dihedral finite group. We add the distance functions:

$$dist^2 = t^2 - x^2 - y^2 + z^2 \qquad\qquad dist^2 = t^2 - x^2 - y^2 + z^2$$

$$+ \qquad\qquad\qquad +$$

$$dist^2 = t^2 - x^2 + y^2 - z^2 \qquad\qquad dist^2 = t^2 - x^2 + y^2 - z^2 \qquad (9.1)$$

$$+ \qquad\qquad\qquad +$$

$$dist^2 = t^2 + x^2 - y^2 - z^2 \qquad\qquad dist^2 = t^2 + x^2 - y^2 - z^2$$

$$- -$$

$$3.dist^2 = 3.t^2 - x^2 - y^2 - z^2 \qquad\qquad 3.dist^2 = 3.t^2 - x^2 - y^2 - z^2$$

The 3's on the left-hand side of these distance functions are merely scaling factors that can be ignored. The 3's in front of the t variable are disposed with by choosing the appropriate units in which to measure time. We have two copies of the distance function of our 4-dimensional space-time. We have a left-chiral copy of the distance function of our 4-dimensional space-time, and we have a right chiral copy of the distance function of our 4-dimensional space-time. Since we have handedness within our 4-dimensional space-time, this seems appropriate. Of course, the distance functions are just real numbers, and so it is meaningless to talk of a chiral distance function.

Rotations:

How about superimposing rotation matrices? We do this by adding the A_3 rotation matrices.

$$
\begin{bmatrix}
\cosh\lambda & \frac{x}{\lambda}\sinh\lambda & \frac{y}{\lambda}\sinh\lambda & \frac{z}{\lambda}\sinh\lambda \\
-\frac{x}{\lambda}\sinh\lambda & \cosh\lambda & -\frac{z}{\lambda}\sinh\lambda & \frac{y}{\lambda}\sinh\lambda \\
\frac{y}{\lambda}\sinh\lambda & -\frac{z}{\lambda}\sinh\lambda & \cosh\lambda & -\frac{x}{\lambda}\sinh\lambda \\
\frac{z}{\lambda}\sinh\lambda & \frac{y}{\lambda}\sinh\lambda & \frac{x}{\lambda}\sinh\lambda & \cosh\lambda
\end{bmatrix}
+
\begin{bmatrix}
\cosh\lambda & \frac{x}{\lambda}\sinh\lambda & \frac{y}{\lambda}\sinh\lambda & \frac{z}{\lambda}\sinh\lambda \\
-\frac{x}{\lambda}\sinh\lambda & \cosh\lambda & \frac{z}{\lambda}\sinh\lambda & -\frac{y}{\lambda}\sinh\lambda \\
\frac{y}{\lambda}\sinh\lambda & \frac{z}{\lambda}\sinh\lambda & \cosh\lambda & \frac{x}{\lambda}\sinh\lambda \\
\frac{z}{\lambda}\sinh\lambda & -\frac{y}{\lambda}\sinh\lambda & -\frac{x}{\lambda}\sinh\lambda & \cosh\lambda
\end{bmatrix}
$$

$$\lambda = \sqrt{-x^2 + y^2 + z^2}$$

$$(9.2)$$

The rotation matrices of the different pairs of A_3 algebras have different angles, and so we form a super-position of only the two rotation matrices of a pair, as in (9.2). We get the super-position rotation matrix:

$$\begin{bmatrix} \cosh \lambda & \dfrac{x}{\lambda}\sinh \lambda & \dfrac{y}{\lambda}\sinh \lambda & \dfrac{z}{\lambda}\sinh \lambda \\[2ex] -\dfrac{x}{\lambda}\sinh \lambda & \cosh \lambda & 0 & 0 \\[2ex] \dfrac{y}{\lambda}\sinh \lambda & 0 & \cosh \lambda & 0 \\[2ex] \dfrac{z}{\lambda}\sinh \lambda & 0 & 0 & \cosh \lambda \end{bmatrix}$$

(9.3)

$$\lambda = \sqrt{-x^2 + y^2 + z^2}$$

This, (9.3), is not a *bona fide* rotation matrix, but it will separate into three $SO(3)$ rotation matrices. We present these three matrices:

$$\begin{bmatrix} \cosh \lambda & \dfrac{x}{\lambda}\sinh \lambda & 0 & 0 \\[2ex] -\dfrac{x}{\lambda}\sinh \lambda & \cosh \lambda & 0 & 0 \\[1ex] 0 & 0 & 1 & 0 \\ 0 & 0 & 0 & 1 \end{bmatrix} = \begin{bmatrix} \cos x & \sin x & 0 & 0 \\ -\sin x & \cos x & 0 & 0 \\ 0 & 0 & 1 & 0 \\ 0 & 0 & 0 & 1 \end{bmatrix}$$

(9.4)

$$\lambda = \sqrt{-x^2} = \hat{i}x \qquad \cosh(\hat{i}x) = \cos x \qquad \sinh(\hat{i}x) = \hat{i}\sin x$$

And:

$$\begin{bmatrix} \cosh y & 0 & \sinh y & 0 \\ 0 & 1 & 0 & 0 \\ \sinh y & 0 & \cosh y & 0 \\ 0 & 0 & 0 & 1 \end{bmatrix} \quad \& \quad \begin{bmatrix} \cosh z & 0 & 0 & \sinh z \\ 0 & 0 & 0 & 0 \\ 0 & 0 & 0 & 0 \\ \sinh z & 0 & 0 & \cosh z \end{bmatrix}$$

(9.5)

$$\lambda = \sqrt{y^2} \qquad\qquad\qquad \lambda = \sqrt{z^2}$$

The distance function of our 4-dimensional space-time is held invariant by these three $SO(3)$ rotations.

This super-position thing seems to work.

We notice that super-positions are not chiral.

But it is not mathematics:

Well, it is not mathematically illegal to add six distance functions because each of them is just a real number. Nor is it mathematically illegal to add two matrices within the same algebra. However, we have to ask why. Why would we want to form these super-positions? There is no real answer to this question that is known to your author. Yes, I can wave my arms around and try to construct some sort of seemingly reasonable justification for taking the super-position, but, ultimately, we can give no sensible justification other than "It seems to produce our universe".

Yes, but I bet this sort of thing happens with other finite group spaces:
No, it does not. The A_3 finite group spaces are unique in this regard among all the infinitude of finite group spaces.

The super-position distance function we have just formed from the six A_3 distance functions is:

$$dist^2 = t^2 - x^2 - y^2 - z^2 \qquad (9.6)$$

This super-position distance function, (9.6), can accommodate six 2-dimensional rotations. It is the only super-position distance function which will allow any kind of rotation. From the infinite number of finite group spaces, we can form an infinite number of super-position distance functions. Only one super-position distance function, the A_3 one, can accommodate any type of rotation – utterly remarkable, but true[31].

In the finite group spaces, we have all possible kinds of rotation. No super-position distance function other than the A_3 super-position distance function, (9.6), will accommodate any of these types of rotation. Do we have a proof of this?

Yes, we do have a proof of this. It is a long laborious proof[32]. The proof has been done by brute-force calculation for all finite groups up to and including order fifteen; this first part of the proof is certainly water-tight. However, the second part of the proof is more theoretical, and there are sporadic finite groups which might not be included by the proof. Perhaps the second part of the proof is not water-tight. Further, a good proof ought not to be long and laborious. There might well be a better proof waiting to be found – it might have already been found[33].

Perhaps it is just coincidence:
Perhaps it is just coincidence, but, if so, then it is one hell of an unlikely coincidence. There is an infinity of super-position distance functions. It is most remarkable that the only one which can support rotation exactly matches our observed 4-dimensional space-time – and do not forget that we also have the Lorentz group $SO(3,1)$ commutation relations within the A_3 algebras.

---

[31] Actually, we can also do this with the two quaternion spaces, but they are of different chirality.
[32] It is a whole book long. See : Dennis Morris & Sophie Lacson : The Uniqueness of our Space-time.
[33] There seems to be a proof of this within the representation theory of finite groups, but it is unclear at this time.

An example:

Let us form the super-position distance function of the two algebraically isomorphic 3-dimensional finite group spaces which are representations of the order six cyclic finite group C_6, (7.24).

$$\exp\left(\begin{bmatrix} a & b & c \\ -c & a & b \\ -b & -c & a \end{bmatrix}\right) \qquad\qquad \exp\left(\begin{bmatrix} a & b & c \\ -c & a & -b \\ -b & c & a \end{bmatrix}\right) \tag{9.7}$$

$$dist^3 = a^3 + b^3 - c^3 + 3abc \qquad\qquad dist^3 = a^3 - b^3 + c^3 + 3abc$$

The super-position distance function is:

$$dist^3 = a^3 + 3abc \tag{9.8}$$

There are no rotations of any dimension which keep invariant this super-position distance function, (9.8), or any part of it.

We cannot fit any 2-dimensional rotations into this 3-dimensional super-position distance function, and nor can we fit any 3-dimensional rotation into this 3-dimensional super-position distance function. We certainly cannot fit any higher dimensional rotations into this 3-dimensional super-position space.

Non-commutative differentiation within our 4-dimensional space-time:

We will differentiate within our 4-dimensional space-time by differentiating within the six A_3 algebras and then taking a super-position of the differentials. However, we must first understand how to differentiate.

Summary:

We have met the concept of super-position.

Our 4-dimensional space-time is the super-position space of the 4-dimensional representations of the order eight dihedral finite group also known as the A_3 algebras.

We will see more examples of super-position later in this book.

In all adventures, there is a 'middle bit' where the adventurer has to struggle forward against a tide of obstacles. It is the success of the adventurer in fighting through the 'middle bit' that enhances the heroism of the adventurer's final success and homecoming.

Chapter 10

Conjugates

In this chapter, we prepare for what comes later by considering conjugation.

Conjugation:
The conjugate of a 2-dimensional Euclidean complex number is a rotation in the reverse direction. We have the 2-dimensional Euclidean complex numbers:

$$Conjugate\left(\begin{bmatrix} r & 0 \\ 0 & r \end{bmatrix}\begin{bmatrix} \cos\theta & \sin\theta \\ -\sin\theta & \cos\theta \end{bmatrix}\right) = \begin{bmatrix} r & 0 \\ 0 & r \end{bmatrix}\begin{bmatrix} \cos(-\theta) & \sin(-\theta) \\ -\sin(-\theta) & \cos(-\theta) \end{bmatrix}$$

(10.1)

$$= \begin{bmatrix} r & 0 \\ 0 & r \end{bmatrix}\begin{bmatrix} \cos\theta & -\sin\theta \\ \sin\theta & \cos\theta \end{bmatrix}$$

Reversing a rotation always rotates the complex number back to the real axis. The result of 'undoing' a rotation is to produce a complex number whose imaginary part is zero and whose real part is the norm, modulus, length[34], of that complex number. The length of the complex number is measured by the distance function of that complex space (finite group space).

Multiplying a complex number by its conjugate always gives the length of the complex number in the form of the distance function of the space:

$$\begin{bmatrix} r & 0 \\ 0 & r \end{bmatrix}\begin{bmatrix} \cos\theta & \sin\theta \\ -\sin\theta & \cos\theta \end{bmatrix}\begin{bmatrix} r & 0 \\ 0 & r \end{bmatrix}\begin{bmatrix} \cos(-\theta) & \sin(-\theta) \\ -\sin(-\theta) & \cos(-\theta) \end{bmatrix}$$

(10.2)

$$= \begin{bmatrix} r^2 & 0 \\ 0 & r^2 \end{bmatrix}\begin{bmatrix} 1 & 0 \\ 0 & 1 \end{bmatrix}$$

And:

$$\begin{bmatrix} x & y \\ -y & x \end{bmatrix}\begin{bmatrix} x & -y \\ y & x \end{bmatrix} = \begin{bmatrix} x^2 + y^2 & 0 \\ 0 & x^2 + y^2 \end{bmatrix}$$

(10.3)

Think distance function in 2-dimensional space is $dist^2 = x^2 + y^2$.

---

[34] Modulus, norm, length, these are three words for the same thing.

Almost the inverse:

We could have normalised everything by appending a real number equal to the squared norm, squared modulus, squared length[35] of the complex number:

$$\frac{\begin{bmatrix} x & y \\ -y & x \end{bmatrix}}{x^2 + y^2} \begin{bmatrix} x & -y \\ y & x \end{bmatrix} = \begin{bmatrix} 1 & 0 \\ 0 & 1 \end{bmatrix} \tag{10.4}$$

This is the inverse, of course. We see that conjugate of a complex number is very closely connected to the inverse of that complex number.

The adjoint matrix:

The adjoint of a matrix is the inverse of the matrix multiplied by the distance function (determinant) of the finite group space of the matrix. Did you notice the word 'inverse' in the previous sentence?

For example, within the 2-dimensional hyperbolic complex numbers, we have:

$$\text{Inverse}\left(\begin{bmatrix} t & z \\ z & t \end{bmatrix} \right) = \frac{1}{t^2 - z^2} \begin{bmatrix} t & -z \\ -z & t \end{bmatrix}$$

$$\text{Adjoint}\left(\begin{bmatrix} t & z \\ z & t \end{bmatrix} \right) = \begin{bmatrix} t & -z \\ -z & t \end{bmatrix} \tag{10.5}$$

$$\begin{bmatrix} t & -z \\ -z & t \end{bmatrix} \begin{bmatrix} t & z \\ z & t \end{bmatrix} = \begin{bmatrix} t^2 - z^2 & 0 \\ 0 & t^2 - z^2 \end{bmatrix}$$

We see that we get the distance function of the finite group space as the product of a matrix and its adjoint.

In the 3-dimensional representation of the finite cyclic group C_3, (3.14), we have:

$$\text{Adjoint}\left(\begin{bmatrix} a & b & c \\ c & a & b \\ b & c & a \end{bmatrix} \right) = \begin{bmatrix} a^2 - bc & c^2 - ab & b^2 - ac \\ b^2 - ac & a^2 - bc & c^2 - ab \\ c^2 - ab & b^2 - ac & a^2 - bc \end{bmatrix} \tag{10.6}$$

$$\begin{bmatrix} a^2 - bc & c^2 - ab & b^2 - ac \\ b^2 - ac & a^2 - bc & c^2 - ab \\ c^2 - ab & b^2 - ac & a^2 - bc \end{bmatrix} \begin{bmatrix} a & b & c \\ c & a & b \\ b & c & a \end{bmatrix} = \begin{bmatrix} a^3 + b^3 + c^3 - 3abc & 0 & 0 \\ 0 & a^3 + b^3 + c^3 - 3abc & 0 \\ 0 & 0 & \sim \end{bmatrix}$$

---

[35] Modulus, norm, length, these are three words for the same thing.

We see that a conjugate is the adjoint. We do not form the conjugate by simply reversing the signs of the imaginary variables, but we almost do. Let us write the conjugate in polar form. A reverse rotation matrix is simply the exponential of the Cartesian form with reversed signs on the imaginary variables and a zero real variable:

$$\exp\left(\begin{bmatrix} 0 & -b & -c \\ -c & 0 & -b \\ -b & -c & 0 \end{bmatrix}\right) = \begin{bmatrix} v_A(-b,-c) & v_B(-b,-c) & v_C(-b,-c) \\ v_C(-b,-c) & v_A(-b,-c) & v_B(-b,-c) \\ v_B(-b,-c) & v_C(-b,-c) & v_A(-b,-c) \end{bmatrix} \quad (10.7)$$

We have:

$$\begin{bmatrix} v_A(-b,-c) & v_B(-b,-c) & v_C(-b,-c) \\ v_C(-b,-c) & v_A(-b,-c) & v_B(-b,-c) \\ v_B(-b,-c) & v_C(-b,-c) & v_A(-b,-c) \end{bmatrix}\begin{bmatrix} v_A(b,c) & v_B(b,c) & v_C(b,c) \\ v_C(b,c) & v_A(b,c) & v_B(b,c) \\ v_B(b,c) & v_C(b,c) & v_A(b,c) \end{bmatrix} = \begin{bmatrix} 1 & 0 & 0 \\ 0 & 1 & 0 \\ 0 & 0 & 1 \end{bmatrix} \quad (10.8)$$

Of course we have this, (10.8), any rotation multiplied by the reverse of that rotation will give the identity. We might think our 3-dimensional polar form conjugate would be:

$$Conj = ? \begin{bmatrix} r & 0 & 0 \\ 0 & r & 0 \\ 0 & 0 & r \end{bmatrix}\begin{bmatrix} v_A(-b,-c) & v_B(-b,-c) & v_C(-b,-c) \\ v_C(-b,-c) & v_A(-b,-c) & v_B(-b,-c) \\ v_B(-b,-c) & v_C(-b,-c) & v_A(-b,-c) \end{bmatrix} \quad (10.9)$$

However, we need to get a cubic radial variable multiplying (10.9) by a similar pair of matrices with a rotation in the opposite direction will give us only a squared radial variable. The proper polar conjugate is:

$$Conj = \begin{bmatrix} r^2 & 0 & 0 \\ 0 & r^2 & 0 \\ 0 & 0 & r^2 \end{bmatrix}\begin{bmatrix} v_A(-b,-c) & v_B(-b,-c) & v_C(-b,-c) \\ v_C(-b,-c) & v_A(-b,-c) & v_B(-b,-c) \\ v_B(-b,-c) & v_C(-b,-c) & v_A(-b,-c) \end{bmatrix} \quad (10.10)$$

The r^2 matrix is the adjoint of the radial matrix in (10.9). Great, but I thought the conjugate is the adjoint. Surely the conjugate is:

$$Conj = Adj\left(\begin{bmatrix} r & 0 & 0 \\ 0 & r & 0 \\ 0 & 0 & r \end{bmatrix}\begin{bmatrix} v_A(b,c) & v_B(b,c) & v_C(b,c) \\ v_C(b,c) & v_A(b,c) & v_B(b,c) \\ v_B(b,c) & v_C(b,c) & v_A(b,c) \end{bmatrix}\right)$$

$$\quad (10.11)$$

$$= \begin{bmatrix} r^2 & 0 & 0 \\ 0 & r^2 & 0 \\ 0 & 0 & r^2 \end{bmatrix}\begin{bmatrix} v_A^2 - v_B v_C & v_C^2 - v_A v_B & v_B^2 - v_A v_C \\ v_B^2 - v_A v_C & v_A^2 - v_B v_C & v_C^2 - v_A v_B \\ v_C^2 - v_A v_B & v_B^2 - v_A v_C & v_A^2 - v_B v_C \end{bmatrix}_{(b,c)}$$

This is also the conjugate. It is a remarkable fact that:

$$v_A(-b,-c) = v_A^2(b,c) - v_B(b,c)v_C(b,c)$$
$$v_B(-b,-c) = v_C^2(b,c) - v_A(b,c)v_B(b,c) \tag{10.12}$$
$$v_C(-b,-c) = v_B^2(b,c) - v_A(b,c)v_C(b,c)$$

Analogous identities to this, (10.12), exist for all kinds of trigonometric functions.

Lovely functions these trigonometric functions! All adventurers meet beautiful maidens or hansom princes somewhere along their journey.

The quaternion conjugate:
We have the left-chiral quaternions:

$$Adj\left(\begin{bmatrix} a & b & c & d \\ -b & a & -d & c \\ -c & d & a & -b \\ -d & -c & b & a \end{bmatrix}\right) = (a^2 + b^2 + c^2 + d^2)\begin{bmatrix} a & -b & -c & -d \\ b & a & d & -c \\ c & -d & a & b \\ d & c & -b & a \end{bmatrix} \tag{10.13}$$

We see that the adjoint is just the quaternion with the signs of the imaginary variables reversed and multiplied by the quaternion distance function.

Obviously, multiplying the adjoint quaternion matrix by the quaternion matrix gives the quaternion distance function:

$$(a^2 + b^2 + c^2 + d^2)\begin{bmatrix} a & -b & -c & -d \\ b & a & d & -c \\ c & -d & a & b \\ d & c & -b & a \end{bmatrix}\begin{bmatrix} a & b & c & d \\ -b & a & -d & c \\ -c & d & a & -b \\ -d & -c & b & a \end{bmatrix} = \begin{bmatrix} (a^2+b^2+c^2+d^2)^2 & 0 & 0 & 0 \\ 0 & \sim & 0 & 0 \\ 0 & 0 & \sim & 0 \\ 0 & 0 & 0 & \sim \end{bmatrix}$$
$$\text{(10.14)}$$

In polar form, the quaternions are, (4.3):

$$\mathbb{H}_{L\chi}^{Rot} = \begin{bmatrix} \cos\lambda & \dfrac{b}{\lambda}\sin\lambda & \dfrac{c}{\lambda}\sin\lambda & \dfrac{d}{\lambda}\sin\lambda \\ -\dfrac{b}{\lambda}\sin\lambda & \cos\lambda & -\dfrac{d}{\lambda}\sin\lambda & \dfrac{c}{\lambda}\sin\lambda \\ -\dfrac{c}{\lambda}\sin\lambda & \dfrac{d}{\lambda}\sin\lambda & \cos\lambda & -\dfrac{b}{\lambda}\sin\lambda \\ -\dfrac{d}{\lambda}\sin\lambda & -\dfrac{c}{\lambda}\sin\lambda & \dfrac{b}{\lambda}\sin\lambda & \cos\lambda \end{bmatrix}_{Clockwise}$$
$$\text{(10.15)}$$

$$\lambda = \sqrt{b^2 + c^2 + d^2}$$

We use 'Clockwise' in a general sense. Reversing the angles gives the reverse rotation:

$$\mathbb{H}_{L\chi}^{Rot} = \begin{bmatrix} \cos\lambda & -\dfrac{b}{\lambda}\sin\lambda & -\dfrac{c}{\lambda}\sin\lambda & -\dfrac{d}{\lambda}\sin\lambda \\[2mm] \dfrac{b}{\lambda}\sin\lambda & \cos\lambda & \dfrac{d}{\lambda}\sin\lambda & -\dfrac{c}{\lambda}\sin\lambda \\[2mm] \dfrac{c}{\lambda}\sin\lambda & -\dfrac{d}{\lambda}\sin\lambda & \cos\lambda & \dfrac{b}{\lambda}\sin\lambda \\[2mm] \dfrac{d}{\lambda}\sin\lambda & \dfrac{c}{\lambda}\sin\lambda & -\dfrac{b}{\lambda}\sin\lambda & \cos\lambda \end{bmatrix}_{Anti-Clockwise}$$

(10.16)

$$\lambda = \sqrt{b^2 + c^2 + d^2}$$

We see that the $\lambda = \sqrt{b^2 + c^2 + d^2}$ is unaffected by the change of sign but the signs of the imaginary variables are reversed. Multiplying (10.15) by (10.16) gives the identity, of course. To get the distance function of quaternion space from such a product, we need to multiply both (10.15) and (10.16) by $(a^2 + b^2 + c^2 + d^2)$; this is just a real number; it can be ignored.

An A$_3$ conjugate:

The A_3 algebras are of interest to us. The A_3 conjugate is very similar to the quaternion conjugate. We have:

$$Adj\left(\begin{bmatrix} a & b & c & d \\ b & a & d & c \\ c & -d & a & -b \\ -d & c & -b & a \end{bmatrix}\right) = (a^2 - b^2 - c^2 + d^2)\begin{bmatrix} a & -b & -c & -d \\ -b & a & -d & -c \\ -c & d & a & b \\ d & -c & b & a \end{bmatrix}$$

(10.17)

All the signs of the imaginary variables are reversed, and, again, we have $(a^2 - b^2 - c^2 + d^2)$ which is the distance function of this finite group space; it is a real number which can be ignored.

Summary:

The product of an algebraic matrix form and the adjoint of that algebraic matrix form is the distance function of the finite group space on the leading diagonal.

The conjugate of an algebraic matrix form is the adjoint of that algebraic matrix form.

Chapter 11

Inner Products

Our exploration now requires us to look at inner-products.

Inner products:

The inner-product is the measure of the 'angle' subtended at the origin between two elements of the finite group space. The measure of the angle is derived as the argument of a trigonometric function. Rotation, angles, and trigonometric functions are bound together by the distance function of the finite group space (division algebra). Rotation holds the particular distance function invariant; the trigonometric functions measure the projections, as measured with the appropriate distance function, on to the axes of a point on the unit circle. Everywhere we look, we see the distance function involved.

Because the distance function is so intimately involved in rotation and angles, we need the inner-product of an element of the finite group space with itself to be that distance function. The inner-product must be based on the adjoint. We have seen this above, (10.3), (10.5), (10.6), & (10.14), when we considered the conjugate.

Of course, what we are calling the conjugate is very closely related to the inner-product. We signify the inner-product with a bold dot, $\bullet$. The inner-product is often called the dot-product.

We take the inner-product of two elements of a finite group space, two matrices, by taking the adjoint of one element and then forming the product of this adjoint and the other element. For example, in the 2-dimensional Euclidean complex numbers, we have the inner-product:

$$\text{Adjoint}\left(\begin{bmatrix} a & b \\ -b & a \end{bmatrix}\right) = \begin{bmatrix} a & -b \\ b & a \end{bmatrix}$$

(11.1)

$$\begin{bmatrix} a & b \\ -b & a \end{bmatrix} \bullet \begin{bmatrix} c & d \\ -d & c \end{bmatrix} = \begin{bmatrix} a & -b \\ b & a \end{bmatrix}\begin{bmatrix} c & d \\ -d & c \end{bmatrix} = \begin{bmatrix} ac+bd & ad-bc \\ -(ad-bc) & ac+bd \end{bmatrix}$$

If the two complex numbers were the same, we would have the distance function as the real part of the inner-product and zero imaginary part.

The imaginary part of (11.1) is equal in magnitude to the traditional cross-product, but it is not a vector sticking out of the complex plane – the complex plane is 2-dimensional.

The dot-product is a measure of the angle between the two complex numbers[36]. In polar form, we have:

$$\begin{bmatrix} \cos\theta & \sin\theta \\ -\sin\theta & \cos\theta \end{bmatrix} \bullet \begin{bmatrix} \cos\phi & \sin\phi \\ -\sin\phi & \cos\phi \end{bmatrix} = \begin{bmatrix} \cos\theta & -\sin\theta \\ \sin\theta & \cos\theta \end{bmatrix} \begin{bmatrix} \cos\phi & \sin\phi \\ -\sin\phi & \cos\phi \end{bmatrix}$$

$$= \begin{bmatrix} \cos\theta\cos\phi + \sin\theta\sin\phi & \cos\theta\sin\phi - \sin\theta\cos\phi \\ -(\cos\theta\sin\phi - \sin\theta\cos\phi) & \cos\theta\cos\phi + \sin\theta\sin\phi \end{bmatrix} \qquad (11.2)$$

$$= \begin{bmatrix} \cos(\theta-\phi) & \sin(\theta-\phi) \\ -\sin(\theta-\phi) & \cos(\theta-\phi) \end{bmatrix}$$

We have been a little deceptive above, (11.1) & (11.2). We used normalised matrices in the polar form, (11.2), and we ought to have used normalised matrices in the Cartesian form. Using normalised Cartesian matrices gives:

$$\frac{\begin{bmatrix} a & b \\ -b & a \end{bmatrix}}{\sqrt{a^2+b^2}} \bullet \frac{\begin{bmatrix} c & d \\ -d & c \end{bmatrix}}{\sqrt{c^2+d^2}} = \frac{\begin{bmatrix} a & -b \\ b & a \end{bmatrix}\begin{bmatrix} c & d \\ -d & c \end{bmatrix}}{\sqrt{a^2+b^2}\sqrt{c^2+d^2}} = \frac{\begin{bmatrix} ac+bd & ad-bc \\ -(ad-bc) & ac+bd \end{bmatrix}}{\sqrt{a^2+b^2}\sqrt{c^2+d^2}} \qquad (11.3)$$

This is the familiar Euclidean inner-product. We have:

$$\begin{bmatrix} \cos(\theta-\phi) & \sin(\theta-\phi) \\ -\sin(\theta-\phi) & \cos(\theta-\phi) \end{bmatrix} = \begin{bmatrix} \dfrac{ac+bd}{\sqrt{a^2+b^2}\sqrt{c^2+d^2}} & \dfrac{ad-bc}{\sqrt{a^2+b^2}\sqrt{c^2+d^2}} \\ -\left(\dfrac{ad-bc}{\sqrt{a^2+b^2}\sqrt{c^2+d^2}}\right) & \dfrac{ac+bd}{\sqrt{a^2+b^2}\sqrt{c^2+d^2}} \end{bmatrix} \qquad (11.4)$$

The inner-product in 2-dimensional space-time:
We have:

$$\text{Adjoint}\left(\begin{bmatrix} t & z \\ z & t \end{bmatrix}\right) = \begin{bmatrix} t & -z \\ -z & t \end{bmatrix}$$

$$(11.5)$$

$$\frac{\begin{bmatrix} t & -z \\ -z & t \end{bmatrix}\begin{bmatrix} s & y \\ y & s \end{bmatrix}}{\sqrt{t^2-z^2}\sqrt{s^2-y^2}} = \frac{\begin{bmatrix} st-yz & ty-sz \\ ty-sz & st-yz \end{bmatrix}}{\sqrt{t^2-z^2}\sqrt{s^2-y^2}}$$

And:

---

[36] Within the complex plane, so also is the cross product a measure of the angle between the two complex numbers, but the angle is measured using the other trigonometric function.

$$\begin{bmatrix} \cosh(-\chi) & \sinh(-\chi) \\ \sinh(-\chi) & \cosh(-\chi) \end{bmatrix}\begin{bmatrix} \cosh\varphi & \sinh\varphi \\ \sinh\varphi & \cosh\varphi \end{bmatrix} = \begin{bmatrix} \cosh\chi & -\sinh\chi \\ -\sinh\chi & \cosh\chi \end{bmatrix}\begin{bmatrix} \cosh\varphi & \sinh\varphi \\ \sinh\varphi & \cosh\varphi \end{bmatrix}$$

(11.6)

$$= \begin{bmatrix} \cosh(\chi-\varphi) & \sinh(\chi-\varphi) \\ \sinh(\chi-\varphi) & \cosh(\chi-\varphi) \end{bmatrix}$$

And:

$$\begin{bmatrix} \cosh(\chi-\varphi) & \sinh(\chi-\varphi) \\ \sinh(\chi-\varphi) & \cosh(\chi-\varphi) \end{bmatrix} = \frac{\begin{bmatrix} st-yz & ty-sz \\ ty-sz & st-yz \end{bmatrix}}{\sqrt{t^2-z^2}\sqrt{s^2-y^2}}$$

(11.7)

We again see that, to obtain the angle between the two elements of 2-dimensional space-time, we have to take the adjoint in (11.5).

The 3-dimensional dot-product:

We will look at the 3-dimensional representation of the order three finite cyclic group C_3. We have the adjoint above, (10.6):

$$\frac{\begin{bmatrix} a & b & c \\ c & a & b \\ b & c & a \end{bmatrix}\bullet\begin{bmatrix} e & f & g \\ g & e & f \\ f & g & e \end{bmatrix}}{\sqrt[3]{a^3+b^3+c^3-3abc}\ \sqrt[3]{e^3+f^3+g^3-3efg}} = \frac{\begin{bmatrix} a^2-bc & c^2-ab & b^2-ac \\ b^2-ac & a^2-bc & c^2-ab \\ c^2-ab & b^2-ac & a^2-bc \end{bmatrix}\begin{bmatrix} e & f & g \\ g & e & f \\ f & g & e \end{bmatrix}}{\sqrt[3]{a^3+b^3+c^3-3abc}\ \sqrt[3]{e^3+f^3+g^3-3efg}}$$

$$= \frac{\begin{bmatrix} a^2e+b^2f+c^2g-bce-acf-abg & \sim & \sim \\ a^2g+b^2e+c^2f-bcf-ace-abf & \sim & \sim \\ a^2f+b^2g+c^2e-bcf-acg-abe & \sim & \sim \end{bmatrix}}{\sqrt[3]{a^3+b^3+c^3-3abc}\ \sqrt[3]{e^3+f^3+g^3-3efg}}$$

(11.8)

We have twiddled out the duplicated elements. This is, of course, a measure of the angle between the two elements of this 3-dimensional finite group space:

$$
\begin{bmatrix} v_A(b,c) & v_B(b,c) & v_C(b,c) \\ v_C(b,c) & v_A(b,c) & v_B(b,c) \\ v_B(b,c) & v_C(b,c) & v_A(b,c) \end{bmatrix} \bullet \begin{bmatrix} v_A(f,g) & v_B(f,g) & v_C(f,g) \\ v_C(f,g) & v_A(f,g) & v_B(f,g) \\ v_B(f,g) & v_C(f,g) & v_A(f,g) \end{bmatrix}
$$

$$
= \begin{bmatrix} v_A(-b,-c) & v_B(-b,-c) & v_C(-b,-c) \\ v_C(-b,-c) & v_A(-b,-c) & v_B(-b,-c) \\ v_B(-b,-c) & v_C(-b,-c) & v_A(-b,-c) \end{bmatrix} \begin{bmatrix} v_A(f,g) & v_B(f,g) & v_C(f,g) \\ v_C(f,g) & v_A(f,g) & v_B(f,g) \\ v_B(f,g) & v_C(f,g) & v_A(f,g) \end{bmatrix} \quad (11.9)
$$

$$
= \begin{bmatrix} v_A\big((f-b),(g-c)\big) & v_B\big((f-b),(g-c)\big) & v_C\big((f-b),(g-c)\big) \\ v_C\big((f-b),(g-c)\big) & v_A\big((f-b),(g-c)\big) & v_B\big((f-b),(g-c)\big) \\ v_B\big((f-b),(g-c)\big) & v_C\big((f-b),(g-c)\big) & v_A\big((f-b),(g-c)\big) \end{bmatrix}
$$

Wherein we have used the trigonometric addition identities of the 3-dimensional nu-functions.[37] We could have used (10.11) and (10.12) instead with the same result.

We have:

$$
\begin{bmatrix} v_A\big((\theta-\phi),(\varphi-\eta)\big) & \sim & \sim \\ v_C\big(((\theta-\phi)),(\varphi-\eta)\big) & \sim & \sim \\ v_B\big(((\theta-\phi)),(\varphi-\eta)\big) & \sim & \sim \end{bmatrix} = \begin{bmatrix} \dfrac{a^2e+b^2f+c^2g-bce-acf-abg}{\sqrt[3]{a^3+b^3+c^3-3abc}\,\sqrt[3]{e^3+f^3+g^3-3efg}} & \sim & \sim \\ \dfrac{a^2g+b^2e+c^2f-bcf-ace-abf}{\sqrt[3]{a^3+b^3+c^3-3abc}\,\sqrt[3]{e^3+f^3+g^3-3efg}} & \sim & \sim \\ \dfrac{a^2f+b^2g+c^2e-bcf-acg-abe}{\sqrt[3]{a^3+b^3+c^3-3abc}\,\sqrt[3]{e^3+f^3+g^3-3efg}} & \sim & \sim \end{bmatrix}
$$
$$(11.10)$$

I love these trigonometric functions – you could not invent them.

Quaternion inner product:

The quaternions are non-commutative. However, the real variable in a product of quaternions is the same regardless of the order of multiplication.

We have given the quaternion adjoint above, (10.13). The Cartesian inner product is:

$$
\frac{\begin{bmatrix} t & -x & -y & -z \\ x & t & z & -y \\ y & -z & t & x \\ z & y & -x & t \end{bmatrix}\begin{bmatrix} p & q & r & s \\ -q & p & -s & r \\ -r & s & p & -q \\ -s & -r & q & p \end{bmatrix}}{\sqrt{t^2+x^2+y^2+z^2}\sqrt{p^2+q^2+r^2+s^2}} = \frac{\begin{bmatrix} tp+xq+yr+zs & \sim & \sim & \sim \\ \sim & & \sim & \sim & \sim \\ \sim & & \sim & \sim & \sim \\ \sim & & \sim & \sim & \sim \end{bmatrix}}{\sqrt{t^2+x^2+y^2+z^2}\sqrt{p^2+q^2+r^2+s^2}} \quad (11.11)
$$

---

[37] See: Dennis Morris : Complex Numbers The Higher Dimensional Forms – 2nd Edition.

This is a measure of the 4-dimensional quaternion angle. It is often presented as being a measure of the 2-dimensional Euclidean angle; looking at the quaternion rotation matrix, (10.15), this error is perhaps understandable.

We have:

$$\lambda = \sqrt{b^2 + c^2 + d^2} \qquad \kappa = \sqrt{f^2 + g^2 + h^2}$$

$$
\begin{bmatrix}
\cos\kappa & -\dfrac{f}{\kappa}\sin\kappa & -\dfrac{g}{\kappa}\sin\kappa & -\dfrac{h}{\kappa}\sin\kappa \\[2mm]
\dfrac{f}{\kappa}\sin\kappa & \cos\kappa & \dfrac{h}{\kappa}\sin\kappa & -\dfrac{g}{\kappa}\sin\kappa \\[2mm]
\dfrac{g}{\kappa}\sin\kappa & -\dfrac{h}{\kappa}\sin\kappa & \cos\kappa & \dfrac{f}{\kappa}\sin\kappa \\[2mm]
\dfrac{h}{\kappa}\sin\kappa & \dfrac{g}{\kappa}\sin\kappa & -\dfrac{f}{\kappa}\sin\kappa & \cos\kappa
\end{bmatrix}
\begin{bmatrix}
\cos\lambda & \dfrac{b}{\lambda}\sin\lambda & \dfrac{c}{\lambda}\sin\lambda & \dfrac{d}{\lambda}\sin\lambda \\[2mm]
-\dfrac{b}{\lambda}\sin\lambda & \cos\lambda & -\dfrac{d}{\lambda}\sin\lambda & \dfrac{c}{\lambda}\sin\lambda \\[2mm]
-\dfrac{c}{\lambda}\sin\lambda & \dfrac{d}{\lambda}\sin\lambda & \cos\lambda & -\dfrac{b}{\lambda}\sin\lambda \\[2mm]
-\dfrac{d}{\lambda}\sin\lambda & -\dfrac{c}{\lambda}\sin\lambda & \dfrac{b}{\lambda}\sin\lambda & \cos\lambda
\end{bmatrix}
$$

$$(11.12)$$

This is:

$$
=
\begin{bmatrix}
\cos\kappa\cos\lambda + \dfrac{(bf+cg+dh)\sin\kappa\sin\lambda}{\kappa\lambda} & \sim & \sim & \sim \\[2mm]
\sim & \sim & \sim & \sim \\[2mm]
\sim & \sim & \sim & \sim \\[2mm]
\sim & \sim & \sim & \sim
\end{bmatrix}
\qquad (11.13)
$$

The correct quaternion inner-product is:

$$\cos\kappa\cos\lambda + \frac{(bf+cg+dh)\sin\kappa\sin\lambda}{\kappa\lambda} = \frac{tp+xq+yr+zs}{\sqrt{t^2+x^2+y^2+z^2}\sqrt{p^2+q^2+r^2+s^2}} \qquad (11.14)$$

Now, within the 2-dimensional Euclidean trigonometric functions, we have:

$$\cos\theta\cos\phi + \sin\theta\sin\phi = \cos(\theta-\phi) \qquad (11.15)$$

But this, (11.15), applies only if:

$$\frac{bf+cg+dh}{\sqrt{b^2+c^2+d^2}\sqrt{f^2+g^2+h^2}} = 1 \qquad (11.16)$$

Which it does not necessarily do, and so we cannot have a simple expression for the angle between two quaternions. On the other hand, there is a quaternion angle between two quaternions and the leading quaternion trigonometric function is a cosine. We therefore have:

$$\cos\left(\sqrt{l^2+m^2+n^2}\right) = \cos\kappa\cos\lambda + \frac{\left(bf+cg+dh\right)\sin\kappa\sin\lambda}{\sqrt{b^2+c^2+d^2}\sqrt{f^2+g^2+h^2}}$$

(11.17)

$$= \frac{tp+xq+yr+zs}{\sqrt{t^2+x^2+y^2+z^2}\sqrt{p^2+q^2+r^2+s^2}}$$

We do have a cosine measuring the inner-product after all.

Summary:

The inner-product within a finite group space is based on the adjoint of one of the two elements in the inner-product.

Chapter 12

Divergences

This chapter is a little out of place in that it refers to differential operators, and we have not yet met a differential operator. We ask the reader to bear with us. We are laying the foundations for the differential operators we will meet shortly.

Bear in mind that, since we are working within finite group spaces, division algebras, the form of the differential operator must match the form of the potential upon which that differential operator acts; this is just multiplicative closure.

The divergence of a field:
Traditionally, in the standard Euclidean vector calculus, the divergence of a field is taken to be the inner product of a differential operator, ∇, and the potential:

$$Div = \nabla \bullet \Phi \tag{12.1}$$

You can find this, (12.1), in any text on vector calculus.

We will accept that a differential operator acts to form an inner product. Indeed, we will base much upon this, (12.1).

We interpret the divergence of a potential to be a measure of how the field of the potential 'spreads out', diverges, into space. Clearly, this 'spreading out' into space will depend upon the nature of the space. 'Spreading out' is different in different spaces. Accepting that the differential operator acts to form an inner product ensures that the potential and the 'spreading out' of the divergence are within the same type of space. Within a particular finite group space, we need both the potential and the differential operator to be of a form that matches the algebraic matrix form of the finite group space, but we also need the divergence, and the curls, to be of the form appropriate to that finite group space[38].

Our acceptance that the differential operator acts to form an inner product as (12.1) means that the differential operator will be based on the adjoint of the potential. I'll repeat that.

The differential operator:
Within a finite group space, the differential operator, including the distribution of signs and variables within the differential operator matches the distribution of signs and variables in the adjoint of the potential. Note the 'within a finite group space' phrase.

---

[38] We will see later that we can form more than one differential operator within a finite group space and thus produce more than one divergence. However, only one of the divergences, the inner product one, 'fits' into the finite group space. The other divergences might 'leak' the field into other spaces.

The distance function of the space:

Taking the distribution of signs within the differential operator to be the same as the distribution of signs within the adjoint of the potential means that the divergence of the field will be of the form of the distance function of the space just as the inner-product is of the form of the distance function of the space.

The traditional Euclidean divergence:

Within the 3-dimensional spatial Euclidean part of our 4-dimensional space-time the divergence matches the distance function of this 3-dimensional spatial part of our 4-dimensional space-time. We have:

$$Div = \nabla \cdot \Phi = \begin{bmatrix} \partial x & \partial y & \partial z \end{bmatrix} \begin{bmatrix} A_x \\ A_y \\ A_z \end{bmatrix} = \frac{\partial A_x}{\partial x} + \frac{\partial A_y}{\partial y} + \frac{\partial A_z}{\partial z}$$

$$dist^2 = \begin{bmatrix} x & y & z \end{bmatrix} \begin{bmatrix} x \\ y \\ z \end{bmatrix} = x^2 + y^2 + z^2 \tag{12.2}$$

$$\begin{bmatrix} x & y & z \end{bmatrix} \cdot \begin{bmatrix} a \\ b \\ c \end{bmatrix} = ax + by + cz$$

This, (12.2), is nonsensical. Inner-products exist in only finite group spaces, but, this is standard notation, and it is something familiar to the reader. We seek to make clear the similarity between the divergence and the distance function (including signs).

In 2-dimension Euclidean space, that is the complex plane, which is a finite group space, we might write the divergence as:

$$\begin{bmatrix} \partial x & -\partial y \\ \partial y & \partial x \end{bmatrix} \begin{bmatrix} f(x,y) & g(x,y) \\ -g(x,y) & f(,y) \end{bmatrix} = \begin{bmatrix} \frac{\partial f}{\partial x} + \frac{\partial g}{\partial y} & \sim \\ \sim & \frac{\partial f}{\partial x} + \frac{\partial g}{\partial y} \end{bmatrix} = \begin{bmatrix} Div & \sim \\ \sim & Div \end{bmatrix} \tag{12.3}$$

Notice that we have used the sign distribution of the adjoint form of the potential for the differential operator. If we are to consistently think of the divergence as an inner-product, dot-product, of a differential operator and a potential, then we must always use the sign distribution of the adjoint form of the potential to determine the sign distribution within the differential operator.

Divergence in our 4-dimensional space-time:

The distance function of our 4-dimensional space-time is:

$$dist^2 = t^2 - x^2 - y^2 - z^2 \tag{12.4}$$

This implies that the divergence within our 4-dimensional space-time is:

$$Div = \frac{\partial A_t}{\partial t} - \frac{\partial A_x}{\partial x} - \frac{\partial A_y}{\partial y} - \frac{\partial A_z}{\partial z} \tag{12.5}$$

The charge of the field:

In classical physics, working in the three spatial dimensions of our 4-dimensional space-time, there is a charge, Q, associated with the divergence corresponding to:

$$\frac{\partial Q}{\partial t} = \frac{\partial A_x}{dx} + \frac{\partial A_y}{dy} + \frac{\partial A_z}{dz}$$

$$\tag{12.6}$$

$$\frac{\partial Q}{\partial t} - \frac{\partial A_x}{dx} - \frac{\partial A_y}{dy} - \frac{\partial A_z}{dz} = 0$$

The charge, Q, is conserved over time if $\frac{\partial Q}{\partial t} = 0$. This is the same as requiring the top line of the above, (12.6), to be equal to zero.

If we come to an expression such as the bottom line of the above, (12.6), as we will, then we speculate that there is a charge identified with the numerator of the $\frac{\partial Q}{\partial t}$ term.

Summary:

We accept within a finite group space the concept of divergences as an inner-product between a differential operator and a potential. This implies that the divergence of a finite group space will match the distance function of that finite group space. This implies that the differential operator must have the same distribution of signs and variables as the adjoint of the potential.

We must match not only the terms of the divergence and the terms of the distance function but we must match also the signs. This will automatically happen if the distribution of variables and signs within the differential operator matches the distribution of signs in the adjoint of the potential.

Phew! You might want to revisit this chapter after reading the next two chapters.

Chapter 13

Differentiation within Finite Group Spaces

We will differentiate within the 2-dimensional Euclidean complex numbers. Of course, the 2-dimensional Euclidean complex numbers are finite group space that is the 2-dimensional representation of the order four cyclic group C_4.

Differentiation within the Euclidean complex numbers:
We seek the differential of the potential:

$$\Phi = \begin{bmatrix} f(x,y) & g(x,y) \\ -g(x,y) & f(x,y) \end{bmatrix} \quad \text{with respect to} \quad \begin{bmatrix} x & y \\ -y & x \end{bmatrix} \tag{13.1}$$

We have:

$$\frac{\partial \begin{bmatrix} f(x,y) & g(x,y) \\ -g(x,y) & f(x,y) \end{bmatrix}}{\partial \begin{bmatrix} x & y \\ -y & x \end{bmatrix}} = \frac{\partial \begin{bmatrix} f(x,y) & g(x,y) \\ -g(x,y) & f(x,y) \end{bmatrix}}{\partial \begin{bmatrix} x & 0 \\ 0 & x \end{bmatrix}} + \frac{\partial \begin{bmatrix} f(x,y) & g(x,y) \\ -g(x,y) & f(x,y) \end{bmatrix}}{\partial \begin{bmatrix} 0 & y \\ -y & 0 \end{bmatrix}} \tag{13.2}$$

This is:

$$\frac{\partial \begin{bmatrix} f(x,y) & 0 \\ 0 & f(x,y) \end{bmatrix}}{\partial \begin{bmatrix} x & 0 \\ 0 & x \end{bmatrix}} + \frac{\partial \begin{bmatrix} 0 & g(x,y) \\ -g(x,y) & 0 \end{bmatrix}}{\partial \begin{bmatrix} x & 0 \\ 0 & x \end{bmatrix}} + \frac{\partial \begin{bmatrix} f(x,y) & 0 \\ 0 & f(x,y) \end{bmatrix}}{\partial \begin{bmatrix} 0 & y \\ -y & 0 \end{bmatrix}} + \frac{\partial \begin{bmatrix} 0 & g(x,y) \\ -g(x,y) & 0 \end{bmatrix}}{\partial \begin{bmatrix} 0 & y \\ -y & 0 \end{bmatrix}}$$

$$(13.3)$$

The left-most of the above four differentials (13.3) is simply a real number function differentiated with respect to a real number. It is of the form:

$$\frac{\partial \begin{bmatrix} f(x,y) & 0 \\ 0 & f(x,y) \end{bmatrix}}{\partial \begin{bmatrix} x & 0 \\ 0 & x \end{bmatrix}} = \begin{bmatrix} \dfrac{\partial f}{\partial x} & 0 \\ 0 & \dfrac{\partial f}{\partial x} \end{bmatrix} \tag{13.4}$$

The second left-most of the above four differentials, (13.3), is an imaginary function differentiated with respect to a real variable. It is of the form:

$$\frac{\partial \begin{bmatrix} 0 & g(x,y) \\ -g(x,y) & 0 \end{bmatrix}}{\partial \begin{bmatrix} x & 0 \\ 0 & x \end{bmatrix}} = \begin{bmatrix} 0 & 1 \\ -1 & 0 \end{bmatrix} \frac{\partial \begin{bmatrix} g(x,y) & 0 \\ 0 & g(x,y) \end{bmatrix}}{\partial \begin{bmatrix} x & 0 \\ 0 & x \end{bmatrix}} = \begin{bmatrix} 0 & 1 \\ -1 & 0 \end{bmatrix} \begin{bmatrix} \frac{\partial g}{\partial x} & 0 \\ 0 & \frac{\partial g}{\partial x} \end{bmatrix} = \begin{bmatrix} 0 & \frac{\partial g}{\partial x} \\ -\frac{\partial g}{\partial x} & 0 \end{bmatrix}$$

(13.5)

The third left-most of the above four differentials, (13.3), is a real function differentiated with respect to an imaginary variable. It is of the form:

$$\frac{\partial \begin{bmatrix} f(x,y) & 0 \\ 0 & f(x,y) \end{bmatrix}}{\partial \begin{bmatrix} 0 & y \\ -y & 0 \end{bmatrix}} = \frac{1}{\begin{bmatrix} 0 & 1 \\ -1 & 0 \end{bmatrix}} \frac{\partial \begin{bmatrix} f(x,y) & 0 \\ 0 & f(x,y) \end{bmatrix}}{\partial \begin{bmatrix} y & 0 \\ 0 & y \end{bmatrix}} = \begin{bmatrix} 0 & -1 \\ 1 & 0 \end{bmatrix} \begin{bmatrix} \frac{\partial f}{\partial y} & 0 \\ 0 & \frac{\partial f}{\partial y} \end{bmatrix} = \begin{bmatrix} 0 & -\frac{\partial f}{\partial y} \\ \frac{\partial f}{\partial y} & 0 \end{bmatrix}$$

(13.6)

Did you notice the change of sign when we took the inverse of the imaginary unit?

The right-most of the above four differentials, (13.3), is an imaginary function differentiated with respect to an imaginary variable. It is of the form:

$$\frac{\partial \begin{bmatrix} 0 & g(x,y) \\ -g(x,y) & 0 \end{bmatrix}}{\partial \begin{bmatrix} 0 & y \\ -y & 0 \end{bmatrix}} = \frac{\begin{bmatrix} 0 & 1 \\ -1 & 0 \end{bmatrix}}{\begin{bmatrix} 0 & 1 \\ -1 & 0 \end{bmatrix}} \frac{\partial \begin{bmatrix} g(x,y) & 0 \\ 0 & g(x,y) \end{bmatrix}}{\partial \begin{bmatrix} y & 0 \\ 0 & y \end{bmatrix}} = \begin{bmatrix} \frac{\partial g}{\partial y} & 0 \\ 0 & \frac{\partial g}{\partial y} \end{bmatrix}$$

(13.7)

The complete differential is the sum of the four separate differentials, (13.4) to (13.7). It is:

$$\frac{\partial \begin{bmatrix} f(x,y) & g(x,y) \\ -g(x,y) & f(x,y) \end{bmatrix}}{\partial \begin{bmatrix} x & y \\ -y & x \end{bmatrix}} = \begin{bmatrix} \frac{\partial f}{\partial x} + \frac{\partial g}{\partial y} & \frac{\partial g}{\partial x} - \frac{\partial f}{\partial y} \\ -\left(\frac{\partial g}{\partial x} - \frac{\partial f}{\partial y}\right) & \frac{\partial f}{\partial x} + \frac{\partial g}{\partial y} \end{bmatrix}$$

(13.8)

If you want the Cauchy-Riemann equations, the essential bits are in the differential matrix.

The reader will recognise this differential as:

$$\frac{\partial \begin{bmatrix} f(x,y) & g(x,y) \\ -g(x,y) & f(x,y) \end{bmatrix}}{\partial \begin{bmatrix} x & y \\ -y & x \end{bmatrix}} = \begin{bmatrix} Div(\Phi) & Curl(\Phi) \\ -Curl(\Phi) & Div(\Phi) \end{bmatrix}$$

(13.9)

What we have just done was calculated independently by your author, but I am told this has been around for many decades.

Did we do it correctly?

From experimental observation, we know that the divergence within 2-dimensional Euclidean space is of the form:

$$Div_{\text{2-dim Euclidean space}} = \frac{\partial f}{\partial x} + \frac{\partial g}{\partial y} \qquad (13.10)$$

This is what we have above, (13.8). We did do it correctly.

An aside:

If we had differentiated a scalar field, we would have:

$$\frac{\partial \begin{bmatrix} f(x,y) & 0 \\ 0 & f(x,y) \end{bmatrix}}{\partial \begin{bmatrix} x & y \\ -y & x \end{bmatrix}} = \begin{bmatrix} \dfrac{\partial f}{\partial x} & -\dfrac{\partial f}{\partial y} \\ \dfrac{\partial f}{\partial y} & \dfrac{\partial f}{\partial x} \end{bmatrix} = grad(\Phi) \qquad (13.11)$$

The differential operator:

Differentiation is a linear operation, and so it ought not to surprise us to find we can use a differential matrix to ease the calculation. We have:

$$\begin{bmatrix} \partial x & -\partial y \\ \partial y & \partial x \end{bmatrix} \begin{bmatrix} f(x,y) & g(x,y) \\ -g(x,y) & f(x,y) \end{bmatrix} = \begin{bmatrix} \dfrac{\partial f}{\partial x} + \dfrac{\partial g}{\partial y} & \dfrac{\partial g}{\partial x} - \dfrac{\partial f}{\partial y} \\ -\left(\dfrac{\partial g}{\partial x} - \dfrac{\partial f}{\partial y}\right) & \dfrac{\partial f}{\partial x} + \dfrac{\partial g}{\partial y} \end{bmatrix} = \begin{bmatrix} Div(\Phi) & Curl(\Phi) \\ -Curl(\Phi) & Div(\Phi) \end{bmatrix}$$

$$(13.12)$$

We call the matrix with the differentials in it the differential operator. The differential operator acts, as if by matrix multiplication, upon the potential to be differentiated.

Notice that, because the variables we differentiate with respect to are in the denominator of the differential, we need to take the inverse of these variables in forming the differential operator. The inverse is just the normalised adjoint.

We really have:

$$\begin{bmatrix} \partial x & \partial y \\ -\partial y & \partial x \end{bmatrix} \bullet \begin{bmatrix} f(x,y) & g(x,y) \\ -g(x,y) & f(x,y) \end{bmatrix} = \begin{bmatrix} Div(\Phi) & Curl(\Phi) \\ -Curl(\Phi) & Div(\Phi) \end{bmatrix}$$

$$(13.13)$$

$$\nabla \bullet \Phi = \begin{bmatrix} Div(\Phi) & Curl(\Phi) \\ -Curl(\Phi) & Div(\Phi) \end{bmatrix}$$

Strictly speaking, there is no such thing as an operator within an algebra. What we have above, (13.12), in the differential operator is no more than an easy way to differentiate properly. In spite of this, we will proceed to use the differential operator because it always works.

Every finite group space has a differential operator. I am told that, within the Euclidean complex numbers, it is known as the 'D' operator.

The signs of the divergence:
What all the shouting and argument is done, the signs we use in the potential are arbitrary. Provided the signs and variables we use in the differential operator match the signs within the adjoint of the potential, we will always get the divergence with signs which match the distance function of the space – we will always get the correct form of the divergence.

The del operator, repetition:
The reader might be familiar with the del operator, ∇, in vector calculus. In vector calculus, we have:

$$Div = \nabla \bullet \Phi \qquad\qquad Curl = \nabla \times \Phi \qquad\qquad (13.14)$$

Within the complex numbers, we might think of the differential operator above, (13.12), as the del operator:

$$\nabla = \begin{bmatrix} \partial x & -\partial y \\ \partial y & \partial x \end{bmatrix} \qquad\qquad (13.15)$$

We see, (13.12), that we get both the del dot-product and the del cross-product in one operator.

Differentiation in the 2-dimensional hyperbolic complex numbers:
Above, we have differentiated within the 2-dimensional Euclidean finite group space, the complex numbers, $\mathbb{C}$, and come to the gradient, divergence, and curl within the 2-dimensional Euclidean finite group space.

Obviously, we can differentiate within the other 2-dimensional finite group space which is 2-dimensional space-time, the hyperbolic complex numbers, $\mathbb{S}$, and obtain a gradient, divergence, and curl within this other 2-dimensional finite group space which is 2-dimensional space-time.

The divergence in 2-dimensional space-time should match the distance function within 2-dimensional space-time:

$$dist^2_{\text{2-dim space-time}} = t^2 - z^2$$

$$(13.16)$$

$$Div_{\text{2-dim space-time}} = \frac{\partial \phi_t}{\partial t} - \frac{\partial A_z}{\partial z}$$

We could achieve this, (13.16), by swapping signs within the differential operator, or by swapping signs in the potential providing that the differential operator is to be of the form of the adjoint of the potential (within the given finite group space).

Declaration:

Within a finite group space, the differential operator will always to be of the form of the adjoint of the potential[39].

This ensures we get a divergence which matches the distance function of the finite group space.

Back to 2-dimensional space-time:

For ease, we use the Cartesian form of the hyperbolic complex numbers algebra, but we remind the reader that only the polar form is an algebra. We will use the differential operator. We have:

$$\Phi_{Space-time} = \begin{bmatrix} f(t,z) & g(t,z) \\ g(t,z) & f(t,z) \end{bmatrix} \tag{13.17}$$

The adjoint of this is:

$$\text{Adjoint}\left(\Phi_{Space-time}\right) = \begin{bmatrix} f(t,z) & -g(t,z) \\ -g(t,z) & f(t,z) \end{bmatrix} \tag{13.18}$$

We copy this to form the differential operator:

$$\text{Adjoint}\left(\begin{bmatrix} \partial t & \partial z \\ \partial z & \partial t \end{bmatrix}\right) = \begin{bmatrix} \partial t & -\partial z \\ -\partial z & \partial t \end{bmatrix} \tag{13.19}$$

$$\text{Diff operator} = d_{2\text{-dim space-time}} = \begin{bmatrix} \partial t & -\partial z \\ -\partial z & \partial t \end{bmatrix}$$

We differentiate as if by matrix multiplication:

$$\begin{bmatrix} \partial t & -\partial z \\ -\partial z & \partial t \end{bmatrix}\begin{bmatrix} \phi(t,z) & g(t,z) \\ g(t,z) & \phi(t,z) \end{bmatrix} = \begin{bmatrix} \dfrac{\partial \phi}{\partial t} - \dfrac{\partial g}{\partial z} & \dfrac{\partial g}{\partial t} - \dfrac{\partial \phi}{\partial z} \\ \dfrac{\partial g}{\partial t} - \dfrac{\partial \phi}{\partial z} & \dfrac{\partial \phi}{\partial t} - \dfrac{\partial g}{\partial z} \end{bmatrix} = \begin{bmatrix} Div & Curl \\ Curl & Div \end{bmatrix}_{2\text{-dim space-time}} \tag{13.20}$$

We see that the space-time divergence and curl are:

---

[39] I think we've repeated this enough by now. In fact, we are going to change this later.

$$\begin{bmatrix} Div\left(\Phi_{Space-time}\right) & Curl\left(\Phi_{Space-time}\right) \\ Curl\left(\Phi_{Space-time}\right) & Div\left(\Phi_{Space-time}\right) \end{bmatrix} = \begin{bmatrix} \dfrac{\partial\phi}{\partial t}-\dfrac{\partial g}{\partial z} & -\dfrac{\partial g}{\partial t}+\dfrac{\partial\phi}{\partial z} \\ -\dfrac{\partial g}{\partial t}+\dfrac{\partial\phi}{\partial z} & \dfrac{\partial\phi}{\partial t}-\dfrac{\partial g}{\partial z} \end{bmatrix} \qquad (13.21)$$

Two 2-dimensional curls:

The curl within 2-dimensional Euclidean space is:

$$Curl^{2-\dim}_{Euclidean} = \frac{\partial g}{\partial x}-\frac{\partial f}{\partial y}$$

$$B = \frac{\partial A_x}{\partial y}-\frac{\partial A_y}{\partial x} \qquad (13.22)$$

Both the differentials are spatial, ∂x & ∂y. We are familiar with this spatial type of curl. The magnetic field is associated with this Euclidean type of curl.

The curl within 2-dimensional space-time is:

$$Curl^{2-\dim}_{Space-time} = -\frac{\partial g}{\partial t}+\frac{\partial\phi}{\partial z}$$

$$E = -\frac{\partial A_z}{\partial t}-\frac{\partial\phi}{\partial z} \qquad (13.23)$$

One of the differentials is time and the other is spatial, ∂t & ∂z. We are also familiar with this type of curl. The electric field is associated with this space-time type of curl; yes, it seems we got a sign wrong, but this is just the arbitrary assignment of direction to a current. Once we have done the calculation, we can simply change the direction of the current by swapping the sign.

The electric field accelerates an electrically charged body in a straight line. Such a change of velocity is a rotation in 2-dimensional space-time.

Look at the position of the curls in the above differential matrices, (13.12) & (13.20):

$$\begin{bmatrix} 0 & Curl\left(\Phi\right) \\ -Curl\left(\Phi\right) & 0 \end{bmatrix}_{Euclidean} \qquad \begin{bmatrix} 0 & Curl\left(\Phi\right) \\ Curl\left(\Phi\right) & 0 \end{bmatrix}_{Space-time} \qquad (13.24)$$

These curls are closely connected to rotation, for example:

$$\exp\left(\begin{bmatrix} 0 & Curl \\ -Curl & 0 \end{bmatrix}\right) = \begin{bmatrix} \cos\left(Curl\right) & \sin\left(Curl\right) \\ -\sin\left(Curl\right) & \cos\left(Curl\right) \end{bmatrix} \qquad (13.25)$$

We see that the curl is an angle of rotation – a phase to use the gauge theory term.

How to differentiate within finite group spaces - 3-dimensional curls:

Above, circa (3.13), we met the 3-dimensional representation of the order three cyclic finite group C_3. Since we are calculating divergences and curls, let us calculate the divergence in this 3-dimensional finite group space. Of course, C_3 is a commutative finite group. We have:

$$C_3 = \begin{bmatrix} a & b & c \\ c & a & b \\ b & c & a \end{bmatrix}$$

(13.26)

First we need the differential operator. We form the differential operator by copying what will be the adjoint of what will be the potential, (10.6). This gives the C_3 differential operator as:

$$\begin{bmatrix} \partial a^2 - \partial b \partial c & \partial c^2 - \partial a \partial b & \partial b^2 - \partial a \partial c \\ \partial b^2 - \partial a \partial c & \partial a^2 - \partial b \partial c & \partial c^2 - \partial a \partial b \\ \partial c^2 - \partial a \partial b & \partial b^2 - \partial a \partial c & \partial a^2 - \partial b \partial c \end{bmatrix}$$

(13.27)

We now need a potential which will give us a divergence which matches the distance function of this space. That distance function is:

$$dist^3 = a^3 + b^3 + c^3 - 3abc$$

(13.28)

Oh, this looks complicated, but it is not that difficult. We have:

$$\Phi_{C_3} = \begin{bmatrix} A_a & A_b & A_c \\ A_c & A_a & A_b \\ A_b & A_c & A_a \end{bmatrix}$$

(13.29)

This, (13.29), is the C_3 potential. We now have:

$$\begin{bmatrix} \partial a^2 - \partial b \partial c & \partial c^2 - \partial a \partial b & \partial b^2 - \partial a \partial c \\ \partial b^2 - \partial a \partial c & \partial a^2 - \partial b \partial c & \partial c^2 - \partial a \partial b \\ \partial c^2 - \partial a \partial b & \partial b^2 - \partial a \partial c & \partial a^2 - \partial b \partial c \end{bmatrix} \begin{bmatrix} A_a & A_b & A_c \\ A_c & A_a & A_b \\ A_b & A_c & A_a \end{bmatrix}$$

(13.30)

$$= \begin{bmatrix} \dfrac{\partial^2 A_a}{\partial a^2} - \dfrac{\partial^2 A_a}{\partial b \partial c} + \dfrac{\partial^2 A_c}{\partial c^2} - \dfrac{\partial^2 A_c}{\partial a \partial b} + \dfrac{\partial^2 A_b}{\partial b^2} - \dfrac{\partial^2 A_b}{\partial a \partial c} & \sim & \sim \\ & \sim & & \\ & \sim & & \sim & \sim \end{bmatrix} = \begin{bmatrix} Div & \sim & \sim \\ \sim & \sim & \sim \\ \sim & \sim & \sim \end{bmatrix}$$

We have twiddled out the duplicate entries and the curls.

We have the correlation between the distance function of this finite group space and the divergence:

$$\dfrac{\partial^2 A_a}{\partial a^2} \quad \dfrac{\partial^2 A_b}{\partial b^2} \quad \dfrac{\partial^2 A_c}{\partial c^2} \quad -\dfrac{\partial^2 A_a}{\partial b \partial c} \quad -\dfrac{\partial^2 A_b}{\partial a \partial c} \quad -\dfrac{\partial^2 A_c}{\partial a \partial b}$$

(13.31)

$$x^3 \qquad y^3 \qquad z^3 \qquad -xyz \qquad -xyz \qquad -xyz$$

Are we doing this correctly? We will never know. We can never find any C_3 fields within our 4-dimensional space-time to test our answer.

This 3-dimensional divergence can never fit into our 4-dimensional space-time any more than a 3-dimensional rotation could fit into our 4-dimensional space-time. We do not expect to see force fields associated with the C_3 potential in our 4-dimensional space-time.

Would you like to see a 4-dimensional C_4 curl based on (2.3)? Do it yourself. Perhaps you would like to see the C_3 curls which we twiddled out above, (13.30). You can do this quite easily; you already have the C_3 differential operator, (13.27).

Gauge theory:
The standard model of particle physics uses gauge theory in which the forces of nature are Lie group potentials[40] with a 'phase' that varies locally from point to point within our 4-dimensional space-time. By 'phase', they mean angle. We can look at gauge theory as being about a curl that varies locally from point to point in our 4-dimensional space-time.

In $SU(2)$ gauge theory, the locally varying 'phase' is associated with the $SU(2)$ commutation relations. The quaternions have the $SU(2)$ commutation relations. If the quaternions have a curl, and they do, then a locally varying quaternion curl could completely replace $SU(2)$ gauge theory – guess where we are heading.

To calculate the quaternion curl, we will need to be able to differentiate within a non-commutative finite group space; remember the commutator is an algebraic operation within the quaternions.

Our 4-dimensional space-time in general:
Curl is associated with rotation. There are only 2-dimensional rotations, both types, in our 4-dimensional space-time, and, therefore only curls of the same nature as the two types of 2-dimensional curls can be manifest in our 4-dimensional space-time.

We have seen that all commutative finite group spaces of dimension greater than two have curls with more than two terms.

Since there is an infinity of commutative finite groups, there is an infinity of different types of curl, but very few of these can be manifest within our 4-dimensional space-time. A curl corresponds to a force field. There can be only very few force fields within our 4-dimensional space-time.

---

[40] I'm presenting this far more directly than it is usually presented.

Summary:

Every commutative finite group space has a differential operator and thus a divergence and a set of curls. We have derived the two curls associated with the two 2-dimensional finite group rotations we find in our 4-dimensional space-time.

The differential operator is of the form of the adjoint of the potential.

A word of warning:

The differential operator is a notational shortcut that gives the correct answer if used only once. You cannot use the differential operator squared to do double differentiation in one step. You must do each step one at a time.

Chapter 14

Non-Commutative Differentiation

Within the quaternions and the A_3 algebras, we have the commutator as a *bona fide* algebraic operation – two permutations in, and one permutation out. The commutator is a multiplicative algebraic operation. Where-ever we have a multiplicative algebraic operation, we can form a ratio - we have a differential.

The differential operator again:

The tedious and expansive derivation of the quaternion differential operators is done in detail elsewhere[41]. We will not repeat those many pages of large matrices here.

In short, we form the differential operator by:

1) Calculate the adjoint matrix of the potential.
2) Substitute a ∂_i in place of every variable i in the adjoint matrix.

The left-chiral quaternions are:

$$\mathbb{H}_{L\chi} = \begin{bmatrix} a & b & c & d \\ -b & a & -d & c \\ -c & d & a & -b \\ -d & -c & b & a \end{bmatrix} \tag{14.1}$$

The adjoint of this, (14.1), is:

$$\mathbb{H}_{L\chi}^{Adjoint} = \begin{bmatrix} a & -b & -c & -d \\ b & a & d & -c \\ c & -d & a & b \\ a & c & -b & a \end{bmatrix} \tag{14.2}$$

The adjoint of this, (14.1), leads to the differential operator:

$$\mathbb{H}_{L\chi}^{Diff-op} = d = \begin{bmatrix} \partial a & -\partial b & -\partial c & -\partial d \\ \partial b & \partial a & \partial d & -\partial c \\ \partial c & -\partial d & \partial a & \partial b \\ \partial a & \partial c & -\partial b & \partial a \end{bmatrix} \tag{14.3}$$

This differential operator will act, as if by matrix multiplication upon a quaternion potential.

---

[41] Dennis Morris : The Physics of Empty Space.

69

Clearly, although we have not stated it before, each separate algebra (there are six separate A_3 algebras and two separate quaternion algebras) has its own differential operator.

Different signs in the potentials:
Suppose we alter the sign of a variable in the potential. Consider:

$$\Phi^{\mathbb{H}_{L\chi}} = \begin{bmatrix} a & b & c & -d \\ -b & a & d & c \\ -c & -d & a & -b \\ d & -c & b & a \end{bmatrix} \tag{14.4}$$

Wherein we have swapped the sign of the d variable. In this case, we take the adjoint of this potential, (14.4), which is:

$$\Phi^{\mathbb{H}_{L\chi}}_{Adjoint} = \begin{bmatrix} a & -b & -c & d \\ b & a & -d & -c \\ c & d & a & b \\ -d & c & -b & a \end{bmatrix} \tag{14.5}$$

And we form the differential operator for this potential based upon this, (14.5), adjoint potential.

$$\mathbb{H}^{Diff-op}_{L\chi} = d = \begin{bmatrix} \partial a & -\partial b & -\partial c & \partial d \\ \partial b & \partial a & -\partial d & -\partial c \\ \partial c & \partial d & \partial a & \partial b \\ -\partial a & \partial c & -\partial b & \partial a \end{bmatrix} \tag{14.6}$$

Notice the sign of the ∂d.

By choosing the differential operator to match the distribution of signs and variables within the adjoint of the potential, we ensure that the divergence will match the distance function of the finite group space in which we are working.

What happens if we dump this worshipping of the adjoint and we choose a differently signed differentiation operator? – wait and see. It is very interesting.

Non-commutative differentiation:
We form the commutator and the anti-commutator using the differential operator, (14.6). We have:

$$E = \frac{1}{2}(d\mathbb{H} + \mathbb{H}d)$$
$$B = \frac{1}{2}(d\mathbb{H} - \mathbb{H}d) \tag{14.7}$$

The d is the appropriate differential operator for the potential of the particular algebra. For example, using the left-chiral quaternions, more expansively, we have:

$$E = \frac{1}{2}\left(\begin{bmatrix} \partial a & -\partial b & -\partial c & -\partial d \\ \partial b & \partial a & \partial d & -\partial c \\ \partial c & -\partial d & \partial a & \partial b \\ \partial a & \partial c & -\partial b & \partial a \end{bmatrix} \begin{bmatrix} a & b & c & d \\ -b & a & -d & c \\ -c & d & a & -b \\ -d & -c & b & a \end{bmatrix} + \begin{bmatrix} a & b & c & d \\ -b & a & -d & c \\ -c & d & a & -b \\ -d & -c & b & a \end{bmatrix} \begin{bmatrix} \partial a & -\partial b & -\partial c & -\partial d \\ \partial b & \partial a & \partial d & -\partial c \\ \partial c & -\partial d & \partial a & \partial b \\ \partial a & \partial c & -\partial b & \partial a \end{bmatrix} \right)$$

(14.8)

And:

$$B = \frac{1}{2}\left(\begin{bmatrix} \partial a & -\partial b & -\partial c & -\partial d \\ \partial b & \partial a & \partial d & -\partial c \\ \partial c & -\partial d & \partial a & \partial b \\ \partial a & \partial c & -\partial b & \partial a \end{bmatrix} \begin{bmatrix} a & b & c & d \\ -b & a & -d & c \\ -c & d & a & -b \\ -d & -c & b & a \end{bmatrix} - \begin{bmatrix} a & b & c & d \\ -b & a & -d & c \\ -c & d & a & -b \\ -d & -c & b & a \end{bmatrix} \begin{bmatrix} \partial a & -\partial b & -\partial c & -\partial d \\ \partial b & \partial a & \partial d & -\partial c \\ \partial c & -\partial d & \partial a & \partial b \\ \partial a & \partial c & -\partial b & \partial a \end{bmatrix} \right)$$

(14.9)

In non-commutative differentiation, we get two differentials; that is we get two fields from one potential. We get two differentials, the E-field or E-differential, and the B-field of B-differential. Using the well-established commutative differentiation gives only one differential.

We call these fields the E-field and the B-field because these seem to correspond, in the quaternion case, to the electric field and the magnetic field of classical electro-magnetism.

There we have it. The above is non-commutative differentiation. It is based on the *bona fide* algebraic operation that is the commutator. The non-commutative differential exists in only algebras, finite group spaces, which have the commutator operation. For example, the 6-dimensional representation space, finite group space, of the non-commutative order six symmetric finite group, S_3, does not hold the commutator operation and so we cannot form a differential within this particular 6-dimensional finite group space.

We can always form a commutative differential within a commutative finite group space. The non-commutative spaces in which we can form a non-commutative differential are much rarer. Certainly, in every case, the n-dimensional representation of an order n non-commutative group does not hold the commutator and so does not have a differential – it does not have a divergence or a curl.

Non-commutative differentiation of commutative potentials:
Let us apply the non-commutative differential operation to a commutative potential. We will use the 2-dimensional representation of the order four cyclic finite group C_4 which we commonly know as the Euclidean complex numbers. We have:

$$\Phi = \begin{bmatrix} \phi(x,y) & A_b(x,y) \\ -A_b(x,y) & \phi(x,y) \end{bmatrix} \qquad d = \begin{bmatrix} \partial x & -\partial y \\ \partial y & \partial x \end{bmatrix}$$

(14.10)

And:

$$E = \frac{1}{2}\left(\begin{bmatrix} \partial x & -\partial y \\ \partial y & \partial x \end{bmatrix} \begin{bmatrix} \phi(x,y) & A_b(x,y) \\ -A_b(x,y) & \phi(x,y) \end{bmatrix} + \begin{bmatrix} \phi(x,y) & A_b(x,y) \\ -A_b(x,y) & \phi(x,y) \end{bmatrix} \begin{bmatrix} \partial x & -\partial y \\ \partial y & \partial x \end{bmatrix} \right)$$

(14.11)

$$E = \begin{bmatrix} \dfrac{\partial \phi}{\partial x} + \dfrac{\partial A_b}{\partial y} & \dfrac{\partial A_b}{\partial x} - \dfrac{\partial \phi}{\partial y} \\ -\left(\dfrac{\partial A_b}{\partial x} - \dfrac{\partial \phi}{\partial y} \right) & \dfrac{\partial \phi}{\partial x} + \dfrac{\partial A_b}{\partial y} \end{bmatrix}$$

This, as we might have expected, is the commutative differential. Clearly, the B-field is zero.

$$B = \frac{1}{2}\left(\begin{bmatrix} \partial x & -\partial y \\ \partial y & \partial x \end{bmatrix} \begin{bmatrix} \phi(x,y) & A_b(x,y) \\ -A_b(x,y) & \phi(x,y) \end{bmatrix} - \begin{bmatrix} \phi(x,y) & A_b(x,y) \\ -A_b(x,y) & \phi(x,y) \end{bmatrix} \begin{bmatrix} \partial x & -\partial y \\ \partial y & \partial x \end{bmatrix} \right)$$

(14.12)

$$B = \begin{bmatrix} 0 & 0 \\ 0 & 0 \end{bmatrix}$$

We see that non-commutative differentiation subsumes within itself the commutative differential. The familiar commutative differential is just the E-field with a zero B-field.

Within the quaternions, the real variable is a wholly commutative variable. If the non-commutative differential did not hold within it the commutative differential, we would be unable to differentiate in the quaternion finite group spaces.

Non-commutativity in 8-dimensions:

There are three types of non-commutative 8-dimensional $C_2 \times C_2 \times C_2$ algebras, but, in addition to the commutative real variable, each 8-dimensional $C_2 \times C_2 \times C_2$ algebra also has an imaginary variable which is commutative. There are also commutative relations between some other imaginary variables. We assert that because commutative differentiation is subsumed by non-commutative differentiation, we can differentiate these algebras even though they are a mixture of commutative imaginary variables and non-commutative imaginary variables.

The history of non-commutative differentiation:

In the 1960's David Hestenes tried to construct the non-commutative differential within Clifford algebras. He kind of succeeded, but, within Clifford algebras, there is the concept that some variables are vectors and some other variables are bi-vectors and some other variables are tri-vectors etc.. Hestenes was forced to introduce the Hodge dual which converts a tri-vector into a vector – it all gets a little messy.

The next step forward was taken by Peter Michael Jack[42] in 2003. Jack used non-commutative differentiation to present the Maxwell equations of classical electromagnetism, unfortunately, Jack used obscure notation and his differentiation was 'plucked out of thin air' simply because it seemed to work.

Your author made the next step in 2013 by using matrix notation to express the quaternions. Suddenly, because of the choice of notation, all the obscurity of Jack's usage disappears. However, at this point, the non-commutative differential is still 'plucked out of thin air'. It was in 2017 that your author showed the commutator to be a *bona fide* algebraic operation and thereby placed the non-commutative differential on a solid basis.

Interestingly, and to give credit where credit is due, Peter Michael Jack cites: C. J. Joly : A Manual of Quaternions : 1905 : Art 57 pp 74-77. Your author is unaware of the nature of Joly's contribution.

---

[42] Peter Michael Jack: Physical Space as a Quaternion Structure 1: Maxwell Equations – A Brief Note. arXiv.math-ph0307038v1 18 Jul 2003.

An Example of Non-commutative Differentiation

We will work within the A_3 algebras. The reader will recall that there are six A_3 algebras, (8.4), (8.5) & (8.6). We will begin with the left-chiral SSA A_3 algebra:

$$SSA_{L\chi} = \exp\left(\begin{bmatrix} t & x & y & z \\ x & t & -z & -y \\ y & z & t & x \\ -z & -y & x & t \end{bmatrix}\right) \tag{15.1}$$

Although the A_3 finite group spaces are division algebras in only their polar forms, it is much easier to work with the Cartesian forms and take the exponential at the end.

The distance function of this finite group space is:

$$\det\left(\begin{bmatrix} t & x & y & z \\ x & t & -z & -y \\ y & z & t & x \\ -z & -y & x & t \end{bmatrix}\right) = \left(t^2 - x^2 - y^2 + z^2\right)^2 \tag{15.2}$$

$$dist^4 = \left(t^2 - x^2 - y^2 + z^2\right)^2$$
$$dist^2 = t^2 - x^2 - y^2 + z^2$$

The adjoint of (15.2) is:

$$\left(t^2 - x^2 - y^2 + z^2\right)\begin{bmatrix} a & -b & -c & -d \\ -b & a & d & c \\ -c & -d & a & -b \\ d & c & -b & a \end{bmatrix} \tag{15.3}$$

The fields of the left-chiral SSA A$_3$ finite group space:

We ignore the $\left(t^2 - x^2 - y^2 + z^2\right)$ which is just a real number, and we get the differential operator of this finite group space by copying the adjoint, (15.3):

$$SSA_{l.\chi}^{\text{Diff op}} = d = \begin{bmatrix} \partial a & -\partial b & -\partial c & -\partial d \\ -\partial b & \partial a & \partial d & \partial c \\ -\partial c & -\partial d & \partial a & -\partial b \\ \partial d & \partial c & -\partial b & \partial a \end{bmatrix}$$ (15.4)

The potential[43] is:

$$Pot = \Phi_{SSA_{l.\chi}} = \begin{bmatrix} \phi(t,x,y,z) & A_x(t,x,y,z) & A_y(t,x,y,z) & A_z(t,x,y,z) \\ A_x(t,x,y,z) & \phi(t,x,y,z) & -A_z(t,x,y,z) & -A_y(t,x,y,z) \\ A_y(t,x,y,z) & A_z(t,x,y,z) & \phi(t,x,y,z) & A_x(t,x,y,z) \\ -A_z(t,x,y,z) & -A_y(t,x,y,z) & A_x(t,x,y,z) & \phi(t,x,y,z) \end{bmatrix}$$ (15.5)

We form the E-field as (14.8).

For presentational ease, we give the top row of the matrix only. The E-field is formed from multiplication and addition within this finite group space, and so it is of the same form as (15.2). The remainder of the matrix can be constructed copying the Cartesian matrix in (15.2).

We have the left-chiral SSA E-field:

$$E_{1,1} = \frac{\partial \phi}{\partial t} - \frac{\partial A_x}{\partial x} - \frac{\partial A_y}{\partial y} + \frac{\partial A_y}{\partial y} = Div_{SSA}$$ (15.6)

As anticipated, and guaranteed by the use of the adjoint to form the differential operator, the divergence matches the distance function of this finite group space. We also have:

$$E_{1,2} = -\frac{\partial \phi}{\partial x} + \frac{\partial A_x}{\partial t}$$

$$E_{1,3} = -\frac{\partial \phi}{\partial y} + \frac{\partial A_y}{\partial t}$$ (15.7)

$$E_{1,4} = -\frac{\partial \phi}{\partial z} + \frac{\partial A_z}{\partial t}$$

Now, this is a 4-dimensional finite group space, and so we might have expected a field with four terms. We see that non-commutative differentiation has produced a field with two terms. Perhaps the reader would like to tediously multiply out the E-field of (15.5) to see why we get two term curls rather than 4-term curls.

This field, (15.7), matches the curl of 2-dimensional space-time, (13.21):

$$Curl_{Space-time}^{2-\text{dim}} = -\frac{\partial g}{\partial t} + \frac{\partial \phi}{\partial z}$$ (15.8)

The complete matrix of the left-chiral E-field is:

---

[43] In case you were wondering, this is an entirely legal kind of pot.

$$E - Field_{SSA_{L\chi}} = \begin{bmatrix} E_{1,1} & E_{1,2} & E_{1,3} & E_{1,4} \\ E_{1,2} & E_{1,1} & -E_{1,4} & -E_{1,3} \\ E_{1,3} & E_{1,4} & E_{1,1} & E_{1,2} \\ -E_{1,4} & -E_{1,3} & E_{1,2} & E_{1,1} \end{bmatrix} \tag{15.9}$$

The SSA B-field is formed as (14.9):

$$B_{1,1} = 0$$

$$B_{1,2} = \frac{\partial A_y}{\partial z} - \frac{\partial A_z}{\partial y}$$

$$B_{1,3} = \frac{\partial A_z}{\partial x} - \frac{\partial A_x}{\partial z} \tag{15.10}$$

$$B_{1,4} = \frac{\partial A_y}{\partial x} - \frac{\partial A_x}{\partial y}$$

This field, (15.10), matches the curl of 2-dimensional Euclidean space, (13.12):

$$Curl_{Euclidean}^{2-dim} = -\frac{\partial g}{\partial x} + \frac{\partial f}{\partial y} \tag{15.11}$$

The complete matrix of the left-chiral B-field is:

$$B - Field_{SSA_{L\chi}} = \begin{bmatrix} 0 & B_{1,2} & B_{1,3} & B_{1,4} \\ B_{1,2} & 0 & -B_{1,4} & -B_{1,3} \\ B_{1,3} & B_{1,4} & 0 & B_{1,2} \\ -B_{1,4} & -B_{1,3} & B_{1,2} & 0 \end{bmatrix} \tag{15.12}$$

We have zero divergence in this field.

These curls are of a nature commensurate with the two types of 2-dimensional rotation. They can be manifest within our 4-dimensional space-time.

The fields of the right-chiral SSA A_3 finite group space:

We will now do with the right-chiral SSA A_3 algebra what we did above with the left-chiral form of this algebra. The potential is, (8.4):

$$Pot = \Phi_{SSA_{R\chi}} = \begin{bmatrix} \phi & A_x & A_y & A_z \\ A_x & \phi & A_z & A_y \\ A_y & -A_z & \phi & -A_x \\ -A_z & A_y & -A_x & \phi \end{bmatrix}_{(t,x,y,z)} \tag{15.13}$$

Taking the adjoint gives the differential operator:

$$SSA_{R\chi}^{\text{Diff op}} = d = \begin{bmatrix} \partial a & -\partial b & -\partial c & -\partial d \\ -\partial b & \partial a & -\partial d & -\partial c \\ -\partial c & \partial d & \partial a & \partial b \\ \partial d & -\partial c & \partial b & \partial a \end{bmatrix} \tag{15.14}$$

The E-field is:

$$E_{1,1} = \frac{\partial \phi}{\partial t} - \frac{\partial A_x}{\partial x} \frac{\partial A_y}{\partial y} + \frac{\partial A_y}{\partial y} = Div_{SSA} \tag{15.15}$$

$$E_{1,2} = -\frac{\partial \phi}{\partial x} + \frac{\partial A_x}{\partial t}$$

$$E_{1,3} = -\frac{\partial \phi}{\partial y} + \frac{\partial A_y}{\partial t} \tag{15.16}$$

$$E_{1,4} = -\frac{\partial \phi}{\partial z} + \frac{\partial A_z}{\partial t}$$

This right-chiral E-field is identical on the top row to the E-field of the left-chiral SSA A_3 finite group space, but the matrix differs in the placement of the minus signs. We have the right-chiral E-field:

$$E - Field_{SSA_{R\chi}} = \begin{bmatrix} E_{1,1} & E_{1,2} & E_{1,3} & E_{1,4} \\ E_{1,2} & E_{1,1} & E_{1,4} & E_{1,3} \\ E_{1,3} & -E_{1,4} & E_{1,1} & -E_{1,2} \\ -E_{1,4} & E_{1,3} & -E_{1,2} & E_{1,1} \end{bmatrix} \tag{15.17}$$

The super-position of these two fields is formed by adding the two E-field matrices, (15.9) and (15.17). That super-position is:

$$E - Field_{SSA}^{\text{Super-position}} = \begin{bmatrix} E_{1,1} & E_{1,2} & E_{1,3} & E_{1,4} \\ E_{1,2} & E_{1,1} & 0 & 0 \\ E_{1,3} & 0 & E_{1,1} & 0 \\ -E_{1,4} & 0 & 0 & E_{1,1} \end{bmatrix} \tag{15.18}$$

We see this super-position E-field, (15.18), is a field without chirality; there are no minus signs in the bottom right-hand 3×3 corner of the matrix.

We effectively have three components of the E-field pointing in three directions. The divergence of both the right-chiral and the left-chiral E-fields is the same, and so the super-position field respects the divergence of the SSA A_3 finite group space, and with that the super-position field respects the distance function of this finite group space.

The right-chiral B-field is:

$$B_{1,1} = 0$$

$$B_{1,2} = -\frac{\partial A_y}{\partial z} + \frac{\partial A_z}{\partial y}$$

$$B_{1,3} = -\frac{\partial A_z}{\partial x} + \frac{\partial A_x}{\partial z}$$

$$B_{1,4} = -\frac{\partial A_y}{\partial x} + \frac{\partial A_x}{\partial y}$$

(15.19)

This right-chiral B-field field, (15.19), is the reverse of the left-chiral B-field, (15.10).

The complete matrix of the right-chiral B-field is:

$$B-Field_{SSA_{R\chi}} = \begin{bmatrix} 0 & B_{1,2} & B_{1,3} & B_{1,4} \\ B_{1,2} & 0 & B_{1,4} & B_{1,3} \\ B_{1,3} & -B_{1,4} & 0 & -B_{1,2} \\ -B_{1,4} & B_{1,3} & -B_{1,2} & 0 \end{bmatrix}$$

(15.20)

The super-position SSA B-field is the sum of the two matrices (15.12) & (15.20). Because the fields are the reverse of each other, we get:

$$B-Field_{L\chi} = \begin{bmatrix} 0 & \frac{\partial A_y}{\partial z} - \frac{\partial A_z}{\partial y} & \frac{\partial A_z}{\partial x} - \frac{\partial A_x}{\partial z} & \frac{\partial A_y}{\partial x} - \frac{\partial A_x}{\partial y} \\ \frac{\partial A_y}{\partial z} - \frac{\partial A_z}{\partial y} & 0 & -\left(\frac{\partial A_y}{\partial x} - \frac{\partial A_x}{\partial y}\right) & -\left(\frac{\partial A_z}{\partial x} - \frac{\partial A_x}{\partial z}\right) \\ \frac{\partial A_z}{\partial x} - \frac{\partial A_x}{\partial z} & \frac{\partial A_y}{\partial x} - \frac{\partial A_x}{\partial y} & 0 & \frac{\partial A_y}{\partial z} - \frac{\partial A_z}{\partial y} \\ -\left(\frac{\partial A_y}{\partial x} - \frac{\partial A_x}{\partial y}\right) & -\left(\frac{\partial A_z}{\partial x} - \frac{\partial A_x}{\partial z}\right) & \frac{\partial A_y}{\partial z} - \frac{\partial A_z}{\partial y} & 0 \end{bmatrix}$$

$$+$$

$$B-Field_{R\chi} = \begin{bmatrix} 0 & -\frac{\partial A_y}{\partial z} + \frac{\partial A_z}{\partial y} & -\frac{\partial A_z}{\partial x} + \frac{\partial A_x}{\partial z} & -\frac{\partial A_y}{\partial x} + \frac{\partial A_x}{\partial y} \\ -\frac{\partial A_y}{\partial z} + \frac{\partial A_z}{\partial y} & 0 & -\frac{\partial A_y}{\partial x} + \frac{\partial A_x}{\partial y} & -\frac{\partial A_z}{\partial x} + \frac{\partial A_x}{\partial z} \\ -\frac{\partial A_z}{\partial x} + \frac{\partial A_x}{\partial z} & -\left(-\frac{\partial A_y}{\partial x} + \frac{\partial A_x}{\partial y}\right) & 0 & -\left(-\frac{\partial A_y}{\partial z} + \frac{\partial A_z}{\partial y}\right) \\ -\left(-\frac{\partial A_y}{\partial x} + \frac{\partial A_x}{\partial y}\right) & -\frac{\partial A_z}{\partial x} + \frac{\partial A_x}{\partial z} & -\left(-\frac{\partial A_y}{\partial z} + \frac{\partial A_z}{\partial y}\right) & 0 \end{bmatrix}$$

(15.21)

This sum is:

$$B-Field_{SSA}^{\text{Super-position}} = \begin{bmatrix} 0 & 0 & 0 & 0 \\ 0 & 0 & -\dfrac{\partial A_y}{\partial x}+\dfrac{\partial A_x}{\partial y} & -\dfrac{\partial A_z}{\partial x}+\dfrac{\partial A_x}{\partial z} \\ 0 & \dfrac{\partial A_y}{\partial x}-\dfrac{\partial A_x}{\partial y} & 0 & \dfrac{\partial A_y}{\partial z}-\dfrac{\partial A_z}{\partial y} \\ 0 & -\dfrac{\partial A_z}{\partial x}+\dfrac{\partial A_x}{\partial z} & \dfrac{\partial A_y}{\partial z}-\dfrac{\partial A_z}{\partial y} & 0 \end{bmatrix} \quad (15.22)$$

Comparing the distribution of minus signs in the super-position B-field, we see that the super-position B-field is a chiral field.

$$B_{SSA}^{\text{Super-position}} = \begin{bmatrix} 0 & 0 & 0 & 0 \\ 0 & 0 & - & - \\ 0 & + & 0 & + \\ 0 & - & + & 0 \end{bmatrix} \qquad B_{L\chi}^{SSA} = \begin{bmatrix} 0 & + & + & + \\ + & 0 & - & - \\ + & + & 0 & + \\ - & - & + & 0 \end{bmatrix} \qquad B_{R\chi}^{SSA} = \begin{bmatrix} 0 & + & + & + \\ + & 0 & + & + \\ + & - & 0 & - \\ - & + & - & 0 \end{bmatrix} \quad (15.23)$$

It is a left-chiral field.

Chiral fields:

There are chiral fields in the observed universe. The most famous chiral field is probably the neutrino field. The magnetic field is also a chiral field although its chirality has been forgotten over the last hundred years[44].

Yes, but we can convert this to a right-chiral field by simply reversing the signs of the A_i in the potential, you argue. Firstly, you would still have a chiral field. Secondly, unless you also reversed the sign on the ϕ in the potential, you would not have the correct expression for the divergence of this finite group space.

An important point:

Now comes an important point. The A_3 finite group spaces, like 2-dimensional space-time, have no additive inverses on the real axis. Remember, we have to accept no additive inverses on the real axis to be rid of zero-divisors and singular matrices. The A_3 finite group spaces could not be a finite group spaces if there were additive inverses on the real axis. This means that you cannot reverse the sign

---

[44] The 'unbalanced' chirality of the magnetic field was a hot topic of discussion during the 19th century. Physicists could not accept that the universe prefers its left-hand over its right-hand, and so they forgot all about the chirality of the magnetic field. In 1932, Carl David Anderson (1905-1991) discovered the positron, and so, it seemed, that the chirality of the magnetic field had been resolved. The chirality of the magnetic field has not been resolved because there is an imbalance between matter and anti-matter in the universe.

of the ϕ within a A_3 finite group space. We speculate that, within the A_3 finite group spaces, $\phi = mass$. It follows that there is no negative mass in the universe – that fits with observation.

Within the quaternion spaces, like 2-dimensional Euclidean space, there are additive inverses on the real axis, and so we can reverse the sign of the ϕ in quaternion space. We speculate that, within the quaternion finite group spaces, $\phi = electric\ charge$. It follows that there can be both negative and positive charge in the universe – that fits with observation.

While we are here:

There are many things that one ought to do during one's short mortal tenure upon this Earth, but surely the most important of these things is to calculate E-fields and B-fields. We give the E-fields and B-fields of the other two pairs of A_3 finite group spaces, (8.5) & (8.6).

The SAS left-chiral fields:

The SAS left-chiral E-field is:

$$E_{1,1} = \frac{\partial \phi}{\partial t} - \frac{\partial A_x}{\partial x} + \frac{\partial A_y}{\partial y} - \frac{\partial A_y}{\partial y} = Div_{SAS} \tag{15.24}$$

This is just two signs different from the divergence of the SSA divergence given above, (15.6).

$$E_{1,2} = -\frac{\partial \phi}{\partial x} + \frac{\partial A_x}{\partial t}$$

$$E_{1,3} = -\frac{\partial \phi}{\partial y} + \frac{\partial A_y}{\partial t} \tag{15.25}$$

$$E_{1,4} = -\frac{\partial \phi}{\partial z} + \frac{\partial A_z}{\partial t}$$

This E-field is identical on the top row to the E-fields of the SSA algebras, but the matrix differs, of course, by a few signs. The SAS left-chiral E-field is:

$$E-Field_{SAS_{Lz}} = \begin{bmatrix} E_{1,1} & E_{1,2} & E_{1,3} & E_{1,4} \\ E_{1,2} & E_{1,1} & E_{1,4} & E_{1,3} \\ -E_{1,3} & E_{1,4} & E_{1,1} & -E_{1,2} \\ E_{1,4} & -E_{1,3} & -E_{1,2} & E_{1,1} \end{bmatrix} \tag{15.26}$$

The SAS left-chiral B-field is:

$$B_{1,1} = 0$$

$$B_{1,2} = \frac{\partial A_y}{\partial z} - \frac{\partial A_z}{\partial y}$$

$$B_{1,3} = -\frac{\partial A_z}{\partial x} + \frac{\partial A_x}{\partial z} \qquad (15.27)$$

$$B_{1,4} = -\frac{\partial A_y}{\partial x} + \frac{\partial A_x}{\partial y}$$

The SAS left-chiral B-field matrix is:

$$B-Field_{SAS_{L\chi}} = \begin{bmatrix} 0 & \frac{\partial A_y}{\partial z} - \frac{\partial A_z}{\partial y} & -\frac{\partial A_z}{\partial x} + \frac{\partial A_x}{\partial z} & -\frac{\partial A_y}{\partial x} + \frac{\partial A_x}{\partial y} \\ \frac{\partial A_y}{\partial z} - \frac{\partial A_z}{\partial y} & 0 & -\frac{\partial A_y}{\partial x} + \frac{\partial A_x}{\partial y} & -\frac{\partial A_z}{\partial x} + \frac{\partial A_x}{\partial z} \\ -\left(-\frac{\partial A_z}{\partial x} + \frac{\partial A_x}{\partial z}\right) & -\frac{\partial A_y}{\partial x} + \frac{\partial A_x}{\partial y} & 0 & -\left(\frac{\partial A_y}{\partial z} - \frac{\partial A_z}{\partial y}\right) \\ -\frac{\partial A_y}{\partial x} + \frac{\partial A_x}{\partial y} & -\left(-\frac{\partial A_z}{\partial x} + \frac{\partial A_x}{\partial z}\right) & -\left(\frac{\partial A_y}{\partial z} - \frac{\partial A_z}{\partial y}\right) & 0 \end{bmatrix}$$

$$(15.28)$$

The SAS right-chiral fields:
The SAS right-chiral E-field is:

$$E_{1,1} = \frac{\partial \phi}{\partial t} - \frac{\partial A_x}{\partial x} + \frac{\partial A_y}{\partial y} - \frac{\partial A_y}{\partial y} = Div_{SAS} \qquad (15.29)$$

$$E_{1,2} = -\frac{\partial \phi}{\partial x} + \frac{\partial A_x}{\partial t}$$

$$E_{1,3} = -\frac{\partial \phi}{\partial y} + \frac{\partial A_y}{\partial t} \qquad (15.30)$$

$$E_{1,4} = -\frac{\partial \phi}{\partial z} + \frac{\partial A_z}{\partial t}$$

This E-field is identical on the E-field of the SAS left-chiral finite group space, but the matrix differs, of course, by a few signs. The SAS right-chiral E-field is:

$$E-Field_{SAS_{R\chi}} = \begin{bmatrix} E_{1,1} & E_{1,2} & E_{1,3} & E_{1,4} \\ E_{1,2} & E_{1,1} & -E_{1,4} & -E_{1,3} \\ -E_{1,3} & -E_{1,4} & E_{1,1} & E_{1,2} \\ E_{1,4} & E_{1,3} & E_{1,2} & E_{1,1} \end{bmatrix} \qquad (15.31)$$

The SAS right-chiral B-field is:

$$B_{1,1} = 0$$

$$B_{1,2} = -\frac{\partial A_y}{\partial z} + \frac{\partial A_z}{\partial y}$$

$$B_{1,3} = \frac{\partial A_z}{\partial x} - \frac{\partial A_x}{\partial z}$$

$$B_{1,4} = \frac{\partial A_y}{\partial x} - \frac{\partial A_x}{\partial y}$$

(15.32)

This is the reverse of the left-chiral SAS B-field. The SAS right-chiral B-field matrix is:

$$B - Field_{SAS_{R\chi}} = \begin{bmatrix} 0 & -\frac{\partial A_y}{\partial z} + \frac{\partial A_z}{\partial y} & \frac{\partial A_z}{\partial x} - \frac{\partial A_x}{\partial z} & \frac{\partial A_y}{\partial x} - \frac{\partial A_x}{\partial y} \\ -\frac{\partial A_y}{\partial z} + \frac{\partial A_z}{\partial y} & 0 & -\left(\frac{\partial A_y}{\partial x} - \frac{\partial A_x}{\partial y}\right) & -\left(\frac{\partial A_z}{\partial x} - \frac{\partial A_x}{\partial z}\right) \\ -\left(\frac{\partial A_z}{\partial x} - \frac{\partial A_x}{\partial z}\right) & -\left(\frac{\partial A_y}{\partial x} - \frac{\partial A_x}{\partial y}\right) & 0 & -\frac{\partial A_y}{\partial z} + \frac{\partial A_z}{\partial y} \\ \frac{\partial A_y}{\partial x} - \frac{\partial A_x}{\partial y} & \frac{\partial A_z}{\partial x} - \frac{\partial A_x}{\partial z} & -\frac{\partial A_y}{\partial z} + \frac{\partial A_z}{\partial y} & 0 \end{bmatrix}$$

(15.33)

The SAS super-position E-field is:

$$E - Field_{SAS}^{\text{Super-position}} = \begin{bmatrix} E_{1,1} & E_{1,2} & E_{1,3} & E_{1,4} \\ E_{1,2} & E_{1,1} & 0 & 0 \\ -E_{1,3} & 0 & E_{1,1} & 0 \\ E_{1,4} & 0 & 0 & E_{1,1} \end{bmatrix}$$

(15.34)

The SAS super-position B-field is:

$$B - Field_{SAS}^{\text{Super-position}} = \begin{bmatrix} 0 & 0 & 0 & 0 \\ 0 & 0 & -\frac{\partial A_y}{\partial x} + \frac{\partial A_x}{\partial y} & -\frac{\partial A_z}{\partial x} + \frac{\partial A_x}{\partial z} \\ 0 & -\frac{\partial A_y}{\partial x} + \frac{\partial A_x}{\partial y} & 0 & -\left(\frac{\partial A_y}{\partial z} - \frac{\partial A_z}{\partial y}\right) \\ 0 & -\left(-\frac{\partial A_z}{\partial x} + \frac{\partial A_x}{\partial z}\right) & -\left(\frac{\partial A_y}{\partial z} - \frac{\partial A_z}{\partial y}\right) & 0 \end{bmatrix}$$

(15.35)

Again, we have a left-chiral field.

The ASS left-chiral fields:

The AAS left-chiral E-field is:

$$E_{1,1} = \frac{\partial \phi}{\partial t} + \frac{\partial A_x}{\partial x} - \frac{\partial A_y}{\partial y} - \frac{\partial A_y}{\partial y} = Div_{AAS} \tag{15.36}$$

This is just two signs different from the divergence of the SSA divergence given above, (15.6).

$$E_{1,2} = -\frac{\partial \phi}{\partial x} + \frac{\partial A_x}{\partial t}$$

$$E_{1,3} = -\frac{\partial \phi}{\partial y} + \frac{\partial A_y}{\partial t} \tag{15.37}$$

$$E_{1,4} = -\frac{\partial \phi}{\partial z} + \frac{\partial A_z}{\partial t}$$

This E-field is identical on the top row to the E-fields of the other A_3 algebras, but the matrix differs, of course, by a few signs. The ASS left-chiral E-field is:

$$E-Field_{ASS_{L\chi}} = \begin{bmatrix} E_{1,1} & E_{1,2} & E_{1,3} & E_{1,4} \\ -E_{1,2} & E_{1,1} & -E_{1,4} & E_{1,3} \\ E_{1,3} & -E_{1,4} & E_{1,1} & -E_{1,2} \\ E_{1,4} & E_{1,3} & E_{1,2} & E_{1,1} \end{bmatrix} \tag{15.38}$$

The ASS left-chiral B-field is:

$$B_{1,1} = 0$$

$$B_{1,2} = -\frac{\partial A_y}{\partial z} + \frac{\partial A_z}{\partial y}$$

$$B_{1,3} = \frac{\partial A_z}{\partial x} - \frac{\partial A_x}{\partial z} \tag{15.39}$$

$$B_{1,4} = -\frac{\partial A_y}{\partial x} + \frac{\partial A_x}{\partial y}$$

The ASS left-chiral B-field matrix is:

$$B - Field_{ASS_{L\chi}} = \begin{bmatrix} 0 & -\dfrac{\partial A_y}{\partial z} + \dfrac{\partial A_z}{\partial y} & \dfrac{\partial A_z}{\partial x} - \dfrac{\partial A_x}{\partial z} & -\dfrac{\partial A_y}{\partial x} + \dfrac{\partial A_x}{\partial y} \\ -\left(-\dfrac{\partial A_y}{\partial z} + \dfrac{\partial A_z}{\partial y}\right) & 0 & -\left(-\dfrac{\partial A_y}{\partial x} + \dfrac{\partial A_x}{\partial y}\right) & \dfrac{\partial A_z}{\partial x} - \dfrac{\partial A_x}{\partial z} \\ \dfrac{\partial A_z}{\partial x} - \dfrac{\partial A_x}{\partial z} & -\left(-\dfrac{\partial A_y}{\partial x} + \dfrac{\partial A_x}{\partial y}\right) & 0 & -\left(-\dfrac{\partial A_y}{\partial z} + \dfrac{\partial A_z}{\partial y}\right) \\ -\dfrac{\partial A_y}{\partial x} + \dfrac{\partial A_x}{\partial y} & \dfrac{\partial A_z}{\partial x} - \dfrac{\partial A_x}{\partial z} & -\dfrac{\partial A_y}{\partial z} + \dfrac{\partial A_z}{\partial y} & 0 \end{bmatrix}$$

$$(15.40)$$

The ASS right-chiral fields:
The ASS right-chiral E-field is:

$$E_{1,1} = \frac{\partial \phi}{\partial t} + \frac{\partial A_x}{\partial x} - \frac{\partial A_y}{\partial y} - \frac{\partial A_y}{\partial y} = Div_{ASS} \qquad (15.41)$$

$$E_{1,2} = -\frac{\partial \phi}{\partial x} + \frac{\partial A_x}{\partial t}$$

$$E_{1,3} = -\frac{\partial \phi}{\partial y} + \frac{\partial A_y}{\partial t} \qquad (15.42)$$

$$E_{1,4} = -\frac{\partial \phi}{\partial z} + \frac{\partial A_z}{\partial t}$$

This E-field is identical on the E-fields of the other A_3 algebras.

The ASS right-chiral E-field is:

$$E - Field_{ASS_{R\chi}} = \begin{bmatrix} E_{1,1} & E_{1,2} & E_{1,3} & E_{1,4} \\ -E_{1,2} & E_{1,1} & E_{1,4} & -E_{1,3} \\ E_{1,3} & E_{1,4} & E_{1,1} & E_{1,2} \\ E_{1,4} & -E_{1,3} & -E_{1,2} & E_{1,1} \end{bmatrix} \qquad (15.43)$$

The ASS right-chiral B-field is:

$$B_{1,1} = 0$$

$$B_{1,2} = \frac{\partial A_y}{\partial z} - \frac{\partial A_z}{\partial y}$$

$$B_{1,3} = -\frac{\partial A_z}{\partial x} + \frac{\partial A_x}{\partial z} \qquad (15.44)$$

$$B_{1,4} = \frac{\partial A_y}{\partial x} - \frac{\partial A_x}{\partial y}$$

This is the reverse of the left-chiral ASS B-field. The ASS right-chiral B-field matrix is:

$$B-Field_{ASS_{R\chi}} = \begin{bmatrix} 0 & \dfrac{\partial A_y}{\partial z} - \dfrac{\partial A_z}{\partial y} & -\dfrac{\partial A_z}{\partial x} + \dfrac{\partial A_x}{\partial z} & \dfrac{\partial A_y}{\partial x} - \dfrac{\partial A_x}{\partial y} \\[3mm] -\left(\dfrac{\partial A_y}{\partial z} - \dfrac{\partial A_z}{\partial y} \right) & 0 & \dfrac{\partial A_y}{\partial x} - \dfrac{\partial A_x}{\partial y} & -\left(-\dfrac{\partial A_z}{\partial x} + \dfrac{\partial A_x}{\partial z} \right) \\[3mm] -\dfrac{\partial A_z}{\partial x} + \dfrac{\partial A_x}{\partial z} & \dfrac{\partial A_y}{\partial x} - \dfrac{\partial A_x}{\partial y} & 0 & \dfrac{\partial A_y}{\partial z} - \dfrac{\partial A_z}{\partial y} \\[3mm] \dfrac{\partial A_y}{\partial x} - \dfrac{\partial A_x}{\partial y} & -\left(-\dfrac{\partial A_z}{\partial x} + \dfrac{\partial A_x}{\partial z} \right) & -\left(\dfrac{\partial A_y}{\partial z} - \dfrac{\partial A_z}{\partial y} \right) & 0 \end{bmatrix}$$

$$(15.45)$$

The ASS super-position E-field is:

$$E-Field_{ASS}^{\text{Super-position}} = \begin{bmatrix} E_{1,1} & E_{1,2} & E_{1,3} & E_{1,4} \\ -E_{1,2} & E_{1,1} & 0 & 0 \\ E_{1,3} & 0 & E_{1,1} & 0 \\ E_{1,4} & 0 & 0 & E_{1,1} \end{bmatrix} \qquad (15.46)$$

The ASS super-position B-field is:

$$B-Field_{ASS}^{\text{Super-position}} = \begin{bmatrix} 0 & 0 & 0 & 0 \\[3mm] 0 & 0 & -\left(-\dfrac{\partial A_y}{\partial x} + \dfrac{\partial A_x}{\partial y} \right) & \dfrac{\partial A_z}{\partial x} - \dfrac{\partial A_x}{\partial z} \\[3mm] 0 & -\left(-\dfrac{\partial A_y}{\partial x} + \dfrac{\partial A_x}{\partial y} \right) & 0 & -\left(-\dfrac{\partial A_y}{\partial z} + \dfrac{\partial A_z}{\partial y} \right) \\[3mm] 0 & \dfrac{\partial A_z}{\partial x} - \dfrac{\partial A_x}{\partial z} & -\dfrac{\partial A_y}{\partial z} + \dfrac{\partial A_z}{\partial y} & 0 \end{bmatrix} \quad (15.47)$$

Again, we have a left-chiral field.

The whole E-field super-position and gravity:

The super-position of all six A_3 E-fields is:

$$\text{All } A_3 \text{ Algebras } E^{\text{Super-position}} = \begin{bmatrix} Div & 3E_{1,2} & 3E_{1,3} & 3E_{1,4} \\ E_{1,2} & Div & 0 & 0 \\ E_{1,3} & 0 & Div & 0 \\ E_{1,4} & 0 & 0 & Div \end{bmatrix}$$

(15.48)

$$Div = 3\frac{\partial \phi}{\partial t} - \frac{\partial A_x}{\partial x} - \frac{\partial A_y}{\partial y} - \frac{\partial A_y}{\partial y}$$

We have an imbalance between the top row and the left-most column. If we are allowed to ignore the 3's, we have a wholly symmetric field with a divergence which matches our 4-dimensional space-time. With such a divergence, we assert that this field could fit into our 4-dimensional space-time. We speculate that the charge of this field is mass, and that this field is gravity.

It is within the nature of the A_3 algebras that they have no additive inverses on the real axis. We interpret this to mean that gravity will always be attractive and never repulsive.

The whole B-field super-position and dark matter:
The super-position of all six A_3 B-fields is:

$$\text{All } A_3 \text{ Algebras } B^{\text{Super-position}} = \begin{bmatrix} 0 & 0 & 0 & 0 \\ 0 & 0 & -\dfrac{\partial A_y}{\partial x} + \dfrac{\partial A_x}{\partial y} & -\dfrac{\partial A_z}{\partial x} + \dfrac{\partial A_x}{\partial z} \\ 0 & \dfrac{\partial A_y}{\partial x} - \dfrac{\partial A_x}{\partial y} & 0 & \dfrac{\partial A_y}{\partial z} - \dfrac{\partial A_z}{\partial y} \\ 0 & \dfrac{\partial A_z}{\partial x} - \dfrac{\partial A_x}{\partial z} & -\dfrac{\partial A_y}{\partial z} + \dfrac{\partial A_z}{\partial y} & 0 \end{bmatrix}$$

(15.49)

We have here, a wholly anti-symmetric chiral field with no divergence. We are tempted to speculate that this is a second type of gravitational field associated with the normal gravitational field in a way similar to how the magnetic field is associated with the electric field. Dark matter, perhaps.

Summary:
The A_3 finite group spaces are the unique set of isomorphic finite group spaces whose super-position can hold rotations. They form the only super-position space with geometric structure, and that super-position space is an exact match for our observed 4-dimensional space-time. The match is in the commutation relations, $SO(3,1)$, in the rotations, and in the distance function.

We have seen two fields, a E-field and a B-field, emerge from the A_3 spaces. What are we doing here?

Chapter 16

Another Example of Non-commutative Differentiation

The reader will recall that there are just two 4-dimensional representations of the order eight quaternion group, (4.2) & (4.10). The reader will recall that one of these two quaternion representations is left-chiral and the other is right-chiral.

In this chapter, we will apply non-commutative differentiation to the two quaternion potentials. The commutation relations of the quaternion potentials is $SU(2)_{L\chi}$ or $SU(2)_{R\chi}$, and so, if you prefer, you can think of this as being the differentiation of a $SU(2)$ potential.

The quaternion potentials:

We begin with two potentials which are the left-chiral quaternion potential and the right-chiral quaternion potential:

$$\Phi_{L\chi}^{\mathbb{H}} = \begin{bmatrix} \phi & A_x & A_y & A_z \\ -A_x & \phi & -A_z & A_y \\ -A_y & A_z & \phi & -A_x \\ -A_z & -A_y & A_x & \phi \end{bmatrix} \qquad \Phi_{R\chi}^{\mathbb{H}} = \begin{bmatrix} \phi & A_x & A_y & A_z \\ -A_x & \phi & A_z & -A_y \\ -A_y & -A_z & \phi & A_x \\ -A_z & A_y & -A_x & \phi \end{bmatrix}$$

(16.1)

$$\phi(t,x,y,z), \quad A_x(t,x,y,z), \quad A_y(t,x,y,z), \quad A_z(t,x,y,z)$$

The quaternion distance function is:

$$\det\left(\begin{bmatrix} \phi & A_x & A_y & A_z \\ -A_x & \phi & -A_z & A_y \\ -A_y & A_z & \phi & -A_x \\ -A_z & -A_y & A_x & \phi \end{bmatrix}\right) = \left(\phi^2 + A_x^2 + A_y^2 + A_z^2\right)^2$$

(16.2)

$$dist^2 = t^2 + x^2 + x^2 + x^2$$

The quaternion differential operators:

We are lucky within the $C_2 \times C_2$ group in that all the variables of the $C_2 \times C_2$ algebraic matrix form are symmetric. It follows that, when we eliminate the parameters to get the 4-dimensional representations of the order eight groups, we have variables which are either symmetric or are anti-symmetric.

Following our standard procedure, we form the differential operators as the adjoint of the potential matrix. The quaternion differential operators are:

$$d_{L\chi}^{\mathbb{H}} = \begin{bmatrix} \partial t & -\partial x & -\partial y & -\partial z \\ \partial x & \partial t & \partial z & -\partial y \\ \partial y & -\partial z & \partial t & \partial x \\ \partial z & \partial y & -\partial x & \partial t \end{bmatrix} \qquad d_{R\chi}^{\mathbb{H}} = \begin{bmatrix} \partial t & -\partial x & -\partial y & -\partial z \\ \partial x & \partial t & -\partial z & \partial y \\ \partial y & \partial z & \partial t & -\partial x \\ \partial z & -\partial y & \partial x & \partial t \end{bmatrix} \qquad (16.3)$$

There are only a few minus signs difference between these two operators as there are only a few minus signs difference between the left-chiral quaternions and the right-chiral quaternions – it is the bottom right-hand 3×3 corners.

The quaternion E-fields:

We differentiate these potentials non-commutatively as above, (14.7), using the right-chiral quaternion differential operator to differentiate the right-chiral quaternion potential and using the left-chiral quaternion differential operator to differentiate the left-chiral quaternion potential.

We will not write out the full differential matrices for reasons of presentation. Each differential matrix is a quaternion of the same chirality as the potential from which it is derived, and so we need to give only the top row of the differential matrix to specify the whole matrix (watch the signs).

The left-chiral E-field differentials are:

$$E_{[1,1]}^{L\chi} = \frac{\partial \phi}{\partial t} + \frac{\partial A_x}{\partial x} + \frac{\partial A_y}{\partial y} + \frac{\partial A_z}{\partial z} = Div_{Quaternion_{L\chi}} \qquad (16.4)$$

$$E_{[1,2]}^{L\chi} = -\frac{\partial \phi}{\partial x} + \frac{\partial A_x}{\partial t} \qquad (16.5)$$

$$E_{[1,3]}^{L\chi} = -\frac{\partial \phi}{\partial y} + \frac{\partial A_y}{\partial t} \qquad (16.6)$$

$$E_{[1,4]}^{L\chi} = -\frac{\partial \phi}{\partial z} + \frac{\partial A_z}{\partial t} \qquad (16.7)$$

The whole left-chiral E-field matrix is:

$$E_{L\chi}^{\mathbb{H}} = \begin{bmatrix} E_{[1,1]} & E_{[1,2]} & E_{[1,3]} & E_{[1,4]} \\ -E_{[1,2]} & E_{[1,1]} & -E_{[1,4]} & E_{[1,3]} \\ -E_{[1,3]} & E_{[1,4]} & E_{[1,1]} & -E_{[1,2]} \\ -E_{[1,4]} & -E_{[1,3]} & E_{[1,2]} & E_{[1,1]} \end{bmatrix} \qquad (16.8)$$

We have the required divergence that matches the distance function of quaternion space.

The three spatial parts of this E-field, (16.5) & (16.6) & (16.7) are curls within 2-dimensional space-time, (13.23).

Curls in 2-dimensional space-time correspond to rotations in 2-dimensional space-time and rotations in 2-dimensional space-time correspond to changes of linear velocity – linear accelerations – in our 4-dimensional space-time. This E-field would accelerate an appropriately charged particle in a particular spatial direction except that the divergence does not match our 4-dimensional space-time.

Is this, (16.4), (16.5), (16.6), & (16.7) the electric field of classical electromagnetism? We think not because this is a chiral field and the classical electric field non-chiral.

The top row of the E-field of the right-chiral quaternion potential, (16.1) is an exact match for the top row of the E-field of the left-chiral quaternion potential:

$$E^t_{R\chi} = \frac{\partial \phi}{\partial t} + \frac{\partial A_x}{\partial x} + \frac{\partial A_y}{\partial y} + \frac{\partial A_z}{\partial z}$$

$$E^x_{R\chi} = -\frac{\partial \phi}{\partial x} + \frac{\partial A_x}{\partial t}$$

$$E^y_{R\chi} = -\frac{\partial \phi}{\partial y} + \frac{\partial A_y}{\partial t} \qquad (16.9)$$

$$E^z_{R\chi} = -\frac{\partial \phi}{\partial z} + \frac{\partial A_z}{\partial t}$$

The whole right-chiral E-field matrix is:

$$E^{\mathbb{H}}_{R\chi} = \begin{bmatrix} E_{[1,1]} & E_{[1,2]} & E_{[1,3]} & E_{[1,4]} \\ -E_{[1,2]} & E_{[1,1]} & E_{[1,4]} & -E_{[1,3]} \\ -E_{[1,3]} & -E_{[1,4]} & E_{[1,1]} & E_{[1,2]} \\ -E_{[1,4]} & E_{[1,3]} & -E_{[1,2]} & E_{[1,1]} \end{bmatrix} \qquad (16.10)$$

However, the bottom right-hand 3×3 corner of the left-chiral E-field matrix is of opposite sign to the bottom right-hand 3×3 corner of the right-chiral E-field matrix.

The quaternion B-fields:
The top row of the B-field of the left-chiral quaternion potential, (16.1), is:

$$B^t_{L\chi} = 0$$

$$B^x_{L\chi} = -\frac{\partial A_z}{\partial y} + \frac{\partial A_y}{\partial z}$$

$$B^y_{L\chi} = -\frac{\partial A_x}{\partial z} + \frac{\partial A_z}{\partial x} \qquad (16.11)$$

$$B^z_{L\chi} = -\frac{\partial A_y}{\partial x} + \frac{\partial A_x}{\partial y}$$

The whole left-chiral B-field matrix is:

$$B_{L\chi}^{\mathbb{H}} = \begin{bmatrix} 0 & -\dfrac{\partial A_z}{\partial y} + \dfrac{\partial A_y}{\partial z} & -\dfrac{\partial A_x}{\partial z} + \dfrac{\partial A_z}{\partial x} & -\dfrac{\partial A_y}{\partial x} + \dfrac{\partial A_x}{\partial y} \\[4mm] -\left(-\dfrac{\partial A_z}{\partial y} + \dfrac{\partial A_y}{\partial z}\right) & 0 & -\left(-\dfrac{\partial A_y}{\partial x} + \dfrac{\partial A_x}{\partial y}\right) & -\dfrac{\partial A_x}{\partial z} + \dfrac{\partial A_z}{\partial x} \\[4mm] -\left(-\dfrac{\partial A_x}{\partial z} + \dfrac{\partial A_z}{\partial x}\right) & -\dfrac{\partial A_y}{\partial x} + \dfrac{\partial A_x}{\partial y} & 0 & -\left(-\dfrac{\partial A_z}{\partial y} + \dfrac{\partial A_y}{\partial z}\right) \\[4mm] -\left(-\dfrac{\partial A_y}{\partial x} + \dfrac{\partial A_x}{\partial y}\right) & -\left(-\dfrac{\partial A_x}{\partial z} + \dfrac{\partial A_z}{\partial x}\right) & -\dfrac{\partial A_z}{\partial y} + \dfrac{\partial A_y}{\partial z} & 0 \end{bmatrix} \quad (16.12)$$

Is this is the magnetic field of classical electromagnetism? Again, we think not – see later. The spatial parts of the B-field are curls in 2-dimensional Euclidean space.

The B-field of the right-chiral quaternions is the reverse, negative, of the B-field of the left-chiral quaternions. We have the top row of the B-field of the right-chiral quaternions:

$$B_{R\chi}^t = 0$$

$$B_{R\chi}^x = -\left(-\frac{\partial A_z}{\partial y} + \frac{\partial A_y}{\partial z}\right)$$

$$B_{R\chi}^y = -\left(-\frac{\partial A_x}{\partial z} + \frac{\partial A_z}{\partial x}\right) \qquad (16.13)$$

$$B_{R\chi}^z = -\left(-\frac{\partial A_y}{\partial x} + \frac{\partial A_x}{\partial y}\right)$$

The whole right-chiral B-field matrix is:

$$B_{R\chi}^{\mathbb{H}} = \begin{bmatrix} 0 & \dfrac{\partial A_z}{\partial y} - \dfrac{\partial A_y}{\partial z} & \dfrac{\partial A_x}{\partial z} - \dfrac{\partial A_z}{\partial x} & \dfrac{\partial A_y}{\partial x} - \dfrac{\partial A_x}{\partial y} \\[4mm] -\left(\dfrac{\partial A_z}{\partial y} - \dfrac{\partial A_y}{\partial z}\right) & 0 & \dfrac{\partial A_y}{\partial x} - \dfrac{\partial A_x}{\partial y} & -\left(\dfrac{\partial A_x}{\partial z} - \dfrac{\partial A_z}{\partial x}\right) \\[4mm] -\left(\dfrac{\partial A_x}{\partial z} - \dfrac{\partial A_z}{\partial x}\right) & -\left(\dfrac{\partial A_y}{\partial x} - \dfrac{\partial A_x}{\partial y}\right) & 0 & \dfrac{\partial A_z}{\partial y} - \dfrac{\partial A_y}{\partial z} \\[4mm] -\left(\dfrac{\partial A_y}{\partial x} - \dfrac{\partial A_x}{\partial y}\right) & \dfrac{\partial A_x}{\partial z} - \dfrac{\partial A_z}{\partial x} & -\left(\dfrac{\partial A_z}{\partial y} - \dfrac{\partial A_y}{\partial z}\right) & 0 \end{bmatrix} \quad (16.14)$$

We notice the zero along the leading diagonal.

The absence of B-field charge:

We see that there is no divergence associated with the B-field. There is no B-field charge. We have two fields, the E-field and the B-field associated with one divergence.

The electron and the superposition of the quaternion E-fields:

Recall that the distribution of minus signs in the bottom right-hand 3×3 corner of a left-chiral quaternion are exactly opposite to the signs in the bottom right-hand 3×3 corner of a right-chiral quaternion. The right-chiral E-field, (16.9), is in the same direction as the left-chiral E-field, (16.4) & (16.5) & (16.6) & (16.7). We have:

$$E_{L\chi} = \begin{bmatrix} E_t & E_x & E_y & E_z \\ -E_x & E_t & -E_z & E_y \\ -E_y & E_z & E_t & -E_x \\ -E_z & -E_y & E_x & E_t \end{bmatrix} \qquad E_{R\chi} = \begin{bmatrix} E_t & E_x & E_y & E_z \\ -E_x & E_t & E_z & -E_y \\ -E_y & -E_z & E_t & E_x \\ -E_z & E_y & -E_x & E_t \end{bmatrix}$$

$$(16.15)$$

The left-handed electron   The right-handed electron

We opine that these two E-fields, (16.15), are the field of the left-handed electron and the field of the right-handed electron[45] which are postulated in electro-weak theory. Here, we are rewriting electro-weak theory.

Taking the super-position of the two electron matrices (adding them) gives:

$$E_{\text{Super-position}}^{\mathbb{H}} = 2 \begin{bmatrix} E_t & E_x & E_y & E_z \\ -E_x & E_t & 0 & 0 \\ -E_y & 0 & E_t & 0 \\ -E_z & 0 & 0 & E_t \end{bmatrix}$$

$$(16.16)$$

The electron we observe

This super-position E-field, (16.16), has no chirality – the chirality is in the minus signs in the bottom right-hand 3×3 corner of the matrix which in this matrix, (16.16), are absent. We opine that this, (16.16), is the field of the electron that we observe in our 4-dimensional space-time.

We opine that the act of taking the super-position has converted the maths into physics. Wow! We do not accept this unquestionably.

Electrons:

We need to clarify the matters above. In modern QFT, quantum field theory, particles, such as the electron, as seen as 'excitations' of the field. They are like a single wave that 'stretches' the field locally. This is like a single wave travelling along a rope or a single wave in a rubber sheet. Waves in ropes and in rubber sheets fade away as their energy is dispersed by friction. There is no friction

---

[45] The very successful electro-weak theory postulates that, outside of our 4-dimensional space-time, electrons are massless. Massless electrons travel at the speed of light. There are two types of electrons; these are spin-up electrons and spin-down electrons. In our 4-dimensional space-time, these two electrons are the same particle seen by two observers moving at different velocities, but, if electrons travel at the speed of light, then we have two differently handed electrons. Electro-weak theory uses the Higgs mechanism to give the electrons mass and thus the two types of electrons are one type of electron because they move at less than the velocity of light.

in the fields of QFT, and so the 'wavelet' does not fade away. Thus, we opine that an electron is described by the quaternion wave equation.

The superposition of the quaternion B-fields:

Recall that the distribution of minus signs in the bottom right-hand 3×3 corner of a left-chiral quaternion are exactly opposite to the signs in the bottom right-hand 3×3 corner of a right-chiral quaternion, but also recall that the right-chiral B-field, (16.13), is the negative of the left-chiral B-field, (16.11). We have:

$$
B_{L\chi} = \begin{bmatrix} 0 & B_x & B_y & B_z \\ -B_x & 0 & -B_z & B_y \\ -B_y & B_z & 0 & -B_x \\ -B_z & -B_y & B_x & 0 \end{bmatrix} \qquad
B_{R\chi} = \begin{bmatrix} 0 & B_x & B_y & B_z \\ -B_x & 0 & B_z & -B_y \\ -B_y & -B_z & 0 & B_x \\ -B_z & B_y & -B_x & 0 \end{bmatrix}
$$

(16.17)

The left-handed B-field   The right-handed B-field

We opine that these two B-fields, (16.17), are the neutrino equivalents of the above left-handed and right-handed electron fields[46]. These handed B-fields are not postulated in electro-weak theory. Here, we are still rewriting electro-weak theory.

Taking the super-position of the two B-field matrices (adding them) gives:

$$
B_{\text{Super-position}}^{\mathbb{H}} = \begin{bmatrix} 0 & 0 & 0 & 0 \\ 0 & 0 & -\left(-\dfrac{\partial A_y}{\partial x}+\dfrac{\partial A_x}{\partial y}\right) & -\dfrac{\partial A_x}{\partial z}+\dfrac{\partial A_z}{\partial x} \\ 0 & -\dfrac{\partial A_y}{\partial x}+\dfrac{\partial A_x}{\partial y} & 0 & -\left(-\dfrac{\partial A_z}{\partial y}+\dfrac{\partial A_y}{\partial z}\right) \\ 0 & -\left(-\dfrac{\partial A_x}{\partial z}+\dfrac{\partial A_z}{\partial x}\right) & -\dfrac{\partial A_z}{\partial y}+\dfrac{\partial A_y}{\partial z} & 0 \end{bmatrix} = \begin{bmatrix} 0 & 0 & 0 & 0 \\ 0 & 0 & -B_z & B_y \\ 0 & B_z & 0 & -B_x \\ 0 & -B_y & B_x & 0 \end{bmatrix}
$$

The left-handed neutrino we observe

(16.18)

This super-position quaternion B-field, (16.18), has left-handed chirality. We opine that this left-chiral field, (16.18), is the field of the left-handed neutrino that we observe in our 4-dimensional space-time.

---

[46] There is a problem of confusion here. Within our 4-dimensional space-time, all neutrinos are left-handed and all anti-neutrinos are right-handed. The primary fields of electro-weak theory are outside of our 4-dimensional space-time, and so the two B-fields shown are not the left-handed neutrino and the right-handed anti-neutrino of common parlance. These fields are something new in particle physics.

Anti-matter:

Suppose, we had chosen the quaternion potentials to be reversed in sign:

$$\Phi_{L\chi}^{\mathbb{H}_{Anti}} = \begin{bmatrix} -\phi & -A_x & -A_y & -A_z \\ A_x & -\phi & A_z & -A_y \\ A_y & -A_z & -\phi & A_x \\ A_z & A_y & -A_x & -\phi \end{bmatrix} \qquad \Phi_{R\chi}^{\mathbb{H}_{Anti}} = \begin{bmatrix} -\phi & -A_x & -A_y & -A_z \\ A_x & -\phi & -A_z & A_y \\ A_y & A_z & -\phi & -A_x \\ A_z & -A_y & A_x & -\phi \end{bmatrix}$$

(16.19)

$$\phi(t,x,y,z), \quad A_x(t,x,y,z), \quad A_y(t,x,y,z), \quad A_z(t,x,y,z)$$

The quaternion distance function is the same:

$$dist^2 = t^2 + x^2 + x^2 + x^2$$

(16.20)

Following our standard procedure, we form the differential operators as the adjoint of the potential matrix. The quaternion differential operators are reversed in sign:

$$d_{L\chi}^{\mathbb{H}_{Anti}} = \begin{bmatrix} -\partial t & \partial x & \partial y & \partial z \\ -\partial x & -\partial t & -\partial z & \partial y \\ -\partial y & \partial z & -\partial t & -\partial x \\ -\partial z & -\partial y & \partial x & -\partial t \end{bmatrix} \qquad d_{R\chi}^{\mathbb{H}_{Anti}} = \begin{bmatrix} -\partial t & \partial x & \partial y & \partial z \\ -\partial x & -\partial t & \partial z & -\partial y \\ -\partial y & -\partial z & -\partial t & \partial x \\ -\partial z & \partial y & -\partial x & -\partial t \end{bmatrix}$$

(16.21)

We get the anti-electron field, the positron field, by reversing every sign, including the sign of the real element, which we think of as the charge, on the leading diagonal, in the potentials, (16.1). This leads to a right-handed anti-neutrino. All of these fields satisfy the relativistic energy momentum equation.[47] The relativistic energy momentum equation is just a quaternion inner product (both chiralities and both positive and negative).

Of course, the quaternions have the $SU(2)$ commutation relations used in electro-weak theory.

This super-position stuff does seem to work.

Our adventure seems to be driven in a particular direction.

---

[47] This is done in great detail in the book : Dennis Morris : The Quaternion Dirac Equation.

Chapter 17

But What If?

But what if I form the differential operator in a way other than as a match for the adjoint of the potential? Can I do this? If I do this, will it make sense? Will a different differential operator have any physical meaning?

The variety of divergences:

We have found several types of 4-dimensional finite group space. We have also found our 4-dimensional space-time as a unique super-position space. We have no interest in the 4-dimensional finite group spaces which derive from the order four cyclic finite group C_4, and we have no interest in the A_1 & A_2 commutative finite group spaces which derive from the order four direct product group $C_2 \times C_2$. We list the 4-dimensional spaces in which we are interested:

Our 4-dim Space-time
$$dist^2 = t^2 - x^2 - y^2 - z^2$$
$$Div = \frac{\partial \phi}{\partial t} - \frac{\partial A_x}{\partial x} - \frac{\partial A_y}{\partial y} - \frac{\partial A_z}{\partial z}$$

The SSA A_3 Spaces
$$dist^2 = t^2 - x^2 - y^2 + z^2$$
$$Div = \frac{\partial \phi}{\partial t} - \frac{\partial A_x}{\partial x} - \frac{\partial A_y}{\partial y} + \frac{\partial A_z}{\partial z}$$

The SAS A_3 Spaces
$$dist^2 = t^2 - x^2 + y^2 - z^2$$
$$Div = \frac{\partial \phi}{\partial t} - \frac{\partial A_x}{\partial x} + \frac{\partial A_y}{\partial y} - \frac{\partial A_z}{\partial z}$$

The ASS A_3 Spaces
$$dist^2 = t^2 + x^2 - y^2 - z^2$$
$$Div = \frac{\partial \phi}{\partial t} + \frac{\partial A_x}{\partial x} - \frac{\partial A_y}{\partial y} - \frac{\partial A_z}{\partial z}$$

The Quaternion Spaces
$$dist^2 = t^2 + x^2 + y^2 + z^2$$
$$Div = \frac{\partial \phi}{\partial t} + \frac{\partial A_x}{\partial x} + \frac{\partial A_y}{\partial y} + \frac{\partial A_z}{\partial z} \tag{17.1}$$

Suppose we were to form a differential operator that produced a divergence like:

$$Div? = \frac{\partial \phi}{\partial t} + \frac{\partial A_x}{\partial x} + \frac{\partial A_y}{\partial y} - \frac{\partial A_z}{\partial z} \tag{17.2}$$

There is no 4-dimensional space with a distance function that matches this divergence. We would have produced something that is meaningless in that it can never be expressed within any of the 4-dimensional spaces that exist. The idea gives one pause, does it not.

Leaky spaces:

Now suppose we form a quaternion differential operator that operates on a quaternion potential, and, as if by mistake, the divergence produced matches the divergence of our 4-dimensional space-time. The idea gives one more pause, does it not.

Let us do this:

We begin with the left-chiral quaternion potential:

$$
\Phi_{L\chi}^{\mathbb{H}} = \begin{bmatrix} \phi & -A_x & -A_y & -A_z \\ A_x & \phi & A_z & -A_y \\ A_y & -A_z & \phi & A_x \\ A_z & A_y & -A_x & \phi \end{bmatrix}_{(t,x,y,z)}
\tag{17.3}
$$

This, (17.3), is certainly a left-chiral quaternion. We can act upon it with a quaternion differential operator. We choose:

$$
d_{L\chi} = \begin{bmatrix} \partial t & -\partial x & -\partial y & -\partial z \\ \partial x & \partial t & \partial z & -\partial y \\ \partial y & -\partial z & \partial t & \partial x \\ \partial z & \partial y & -\partial x & \partial t \end{bmatrix}
\tag{17.4}
$$

This, (17.4), is certainly a quaternion differential operator; it is of the form of a quaternion. There is no algebraic reason these two objects, (17.3) & (17.4), cannot be 'multiplied' together.

Without even doing the calculation, we see that the divergence will match the divergence of our 4-dimensional space-time:

$$
d_{L\chi}\Phi_{L\chi}^{\mathbb{H}} = \begin{bmatrix} \partial t & -\partial x & -\partial y & -\partial z \\ \partial x & \partial t & \partial z & -\partial y \\ \partial y & -\partial z & \partial t & \partial x \\ \partial z & \partial y & -\partial x & \partial t \end{bmatrix}\begin{bmatrix} \phi & -A_x & -A_y & -A_z \\ A_x & \phi & A_z & -A_y \\ A_y & -A_z & \phi & A_x \\ A_z & A_y & -A_x & \phi \end{bmatrix} = \begin{bmatrix} \frac{\partial\phi}{\partial t}-\frac{\partial A_x}{\partial x}-\frac{\partial A_y}{\partial y}-\frac{\partial A_z}{\partial z} & \sim & \sim & \sim \\ \sim & & \sim & \sim & \sim \\ \sim & & & \sim & \sim & \sim \\ \sim & & \sim & \sim & \sim \end{bmatrix}
$$

$$
\tag{17.5}
$$

The quaternion field has 'leaked' into our 4-dimensional space-time.

We must expect surprises when we go on adventures.

Is insanity among us?

Leaky spaces? Surely your author is insane, but both (17.3) & (17.4) are within the left-chiral quaternion algebra. They can be multiplied together. The product is a legitimate algebraic entity in that it is a set of permutations. Mathematically, everything is sound, and, after all, the direction we assign to the parts of the potential is entirely arbitrary.

There is much in physics which seems insane. The concept of a cat being both dead and alive at the same time until it is observed is surely insane, but it has come to be an accepted part of physics. The multiverse which postulates every billionth of a second billions of new universes 'spring' into existence is certainly insane, but physics accepts the possibility of it. Until we discovered them, black holes were insane. Even special relativity was deemed insane by those who insanely believed in the ether. There is no shortage of insane mathematicians, and there are even more insane physicists. Thank goodness that your author is not one of them. The reader is encouraged to form their own opinion in this regard.

We reconsider this concept in a later chapter.

The fermion content of the universe:

How many of the finite group spaces have a distance function such that they might 'leak' into our 4-dimensional space-time? Not many. Perhaps the number will exactly match the number of forces within our 4-dimensional space-time universe. You see where we are going.

The leaky quaternion field E-field:

Using the above left-chiral quaternion potential, (17.3), and the above left-chiral quaternion differential operator, (17.4), we get the left-chiral quaternion E-field:

$$E_{[1,1]}^{L\chi} = \frac{\partial \phi}{\partial t} - \frac{\partial A_x}{\partial x} - \frac{\partial A_y}{\partial y} - \frac{\partial A_z}{\partial z} = Div_{\text{4-dim Space-time}} \qquad (17.6)$$

$$E_{[1,2]}^{L\chi} = -\frac{\partial \phi}{\partial x} - \frac{\partial A_x}{\partial t}$$

$$E_{[1,3]}^{L\chi} = -\frac{\partial \phi}{\partial y} - \frac{\partial A_y}{\partial t} \qquad (17.7)$$

$$E_{[1,4]}^{L\chi} = -\frac{\partial \phi}{\partial z} - \frac{\partial A_z}{\partial t}$$

Hm! This, (17.7), matches the classical definition of an electric field.

The whole left-chiral E-field matrix is a chiral field:

$$E_{L\chi}^{\mathbb{H}} = \begin{bmatrix} E_{[1,1]} & E_{[1,2]} & E_{[1,3]} & E_{[1,4]} \\ -E_{[1,2]} & E_{[1,1]} & -E_{[1,4]} & E_{[1,3]} \\ -E_{[1,3]} & E_{[1,4]} & E_{[1,1]} & -E_{[1,2]} \\ -E_{[1,4]} & -E_{[1,3]} & E_{[1,2]} & E_{[1,1]} \end{bmatrix} \tag{17.8}$$

We do the same with the right-chiral quaternions:

$$\Phi_{R\chi}^{\mathbb{H}} = \begin{bmatrix} \phi & -A_x & -A_y & -A_z \\ A_x & \phi & -A_z & A_y \\ A_y & A_z & \phi & -A_x \\ A_z & -A_y & A_x & \phi \end{bmatrix}_{(t,x,y,z)} \qquad d_{R\chi} = \begin{bmatrix} \partial t & -\partial x & -\partial y & -\partial z \\ \partial x & \partial t & -\partial z & \partial y \\ \partial y & \partial z & \partial t & -\partial x \\ \partial z & -\partial y & \partial x & \partial t \end{bmatrix} \tag{17.9}$$

The right-chiral E-fields are:

$$E_{[1,1]}^{R\chi} = \frac{\partial \phi}{\partial t} - \frac{\partial A_x}{\partial x} - \frac{\partial A_y}{\partial y} - \frac{\partial A_z}{\partial z} = Div_{\text{4-dim Space-time}} \tag{17.10}$$

$$E_{[1,2]}^{R\chi} = -\frac{\partial \phi}{\partial x} - \frac{\partial A_x}{\partial t}$$

$$E_{[1,3]}^{R\chi} = -\frac{\partial \phi}{\partial y} - \frac{\partial A_y}{\partial t} \tag{17.11}$$

$$E_{[1,4]}^{R\chi} = -\frac{\partial \phi}{\partial z} - \frac{\partial A_z}{\partial t}$$

These are identical to the left-chiral E-fields. The whole right-chiral E-field matrix is a chiral field:

$$E_{R\chi}^{\mathbb{H}} = \begin{bmatrix} E_{[1,1]} & E_{[1,2]} & E_{[1,3]} & E_{[1,4]} \\ -E_{[1,2]} & E_{[1,1]} & E_{[1,4]} & -E_{[1,3]} \\ -E_{[1,3]} & -E_{[1,4]} & E_{[1,1]} & E_{[1,2]} \\ -E_{[1,4]} & E_{[1,3]} & -E_{[1,2]} & E_{[1,1]} \end{bmatrix} \tag{17.12}$$

We form the super-position of these two E-fields by adding the two matrices:

$$E_{Super-position}^{\mathbb{H}} = \begin{bmatrix} E_{[1,1]} & E_{[1,2]} & E_{[1,3]} & E_{[1,4]} \\ -E_{[1,2]} & E_{[1,1]} & 0 & 0 \\ -E_{[1,3]} & 0 & E_{[1,1]} & 0 \\ -E_{[1,4]} & 0 & 0 & E_{[1,1]} \end{bmatrix} \tag{17.13}$$

$$E_{[1,1]}^{\mathbb{H}} = \frac{\partial \phi}{\partial t} - \frac{\partial A_x}{\partial x} - \frac{\partial A_y}{\partial y} - \frac{\partial A_z}{\partial z} = Div_{\text{4-dim Space-time}}$$

We have a non-chiral E-field in our 4-dimensional space-time.

The leaky quaternion field B-field:
The left-chiral B-field is:

$$B^t_{L\chi} = 0$$

$$B^x_{L\chi} = \frac{\partial A_z}{\partial y} - \frac{\partial A_y}{\partial z}$$

$$B^y_{L\chi} = \frac{\partial A_x}{\partial z} - \frac{\partial A_z}{\partial x}$$

$$B^z_{L\chi} = \frac{\partial A_y}{\partial x} - \frac{\partial A_x}{\partial y}$$

(17.14)

The whole left-chiral B-matrix is:

$$B^{\mathbb{H}}_{L\chi} = \begin{bmatrix} 0 & \frac{\partial A_z}{\partial y} - \frac{\partial A_y}{\partial z} & \frac{\partial A_x}{\partial z} - \frac{\partial A_z}{\partial x} & \frac{\partial A_y}{\partial x} - \frac{\partial A_x}{\partial y} \\ -\left(\frac{\partial A_z}{\partial y} - \frac{\partial A_y}{\partial z}\right) & 0 & -\left(\frac{\partial A_y}{\partial x} - \frac{\partial A_x}{\partial y}\right) & \frac{\partial A_x}{\partial z} - \frac{\partial A_z}{\partial x} \\ -\left(\frac{\partial A_x}{\partial z} - \frac{\partial A_z}{\partial x}\right) & \frac{\partial A_y}{\partial x} - \frac{\partial A_x}{\partial y} & 0 & -\left(\frac{\partial A_z}{\partial y} - \frac{\partial A_y}{\partial z}\right) \\ -\left(\frac{\partial A_y}{\partial x} - \frac{\partial A_x}{\partial y}\right) & -\left(\frac{\partial A_x}{\partial z} - \frac{\partial A_z}{\partial x}\right) & \frac{\partial A_z}{\partial y} - \frac{\partial A_y}{\partial z} & 0 \end{bmatrix}$$

(17.15)

The right-chiral B-field is:

$$B^t_{R\chi} = 0$$

$$B^x_{R\chi} = -\frac{\partial A_z}{\partial y} + \frac{\partial A_y}{\partial z}$$

$$B^y_{R\chi} = -\frac{\partial A_x}{\partial z} + \frac{\partial A_z}{\partial x}$$

$$B^z_{R\chi} = -\frac{\partial A_y}{\partial x} + \frac{\partial A_x}{\partial y}$$

(17.16)

The whole right-chiral B-matrix is:

$$B_{R\chi}^{\mathbb{H}} = \begin{bmatrix} 0 & -\dfrac{\partial A_z}{\partial y} + \dfrac{\partial A_y}{\partial z} & -\dfrac{\partial A_x}{\partial z} + \dfrac{\partial A_z}{\partial x} & -\dfrac{\partial A_y}{\partial x} + \dfrac{\partial A_x}{\partial y} \\[2ex] -\left(-\dfrac{\partial A_z}{\partial y} + \dfrac{\partial A_y}{\partial z}\right) & 0 & -\dfrac{\partial A_y}{\partial x} + \dfrac{\partial A_x}{\partial y} & -\left(-\dfrac{\partial A_x}{\partial z} + \dfrac{\partial A_z}{\partial x}\right) \\[2ex] -\left(-\dfrac{\partial A_x}{\partial z} + \dfrac{\partial A_z}{\partial x}\right) & -\left(-\dfrac{\partial A_y}{\partial x} + \dfrac{\partial A_x}{\partial y}\right) & 0 & -\dfrac{\partial A_z}{\partial y} + \dfrac{\partial A_y}{\partial z} \\[2ex] -\left(-\dfrac{\partial A_y}{\partial x} + \dfrac{\partial A_x}{\partial y}\right) & -\dfrac{\partial A_x}{\partial z} + \dfrac{\partial A_z}{\partial x} & -\left(-\dfrac{\partial A_z}{\partial y} + \dfrac{\partial A_y}{\partial z}\right) & 0 \end{bmatrix} \quad (17.17)$$

The super-position of the right-chiral B-matrix and the left-chiral B-matrix is their sum. This is:

$$B_{Super-position}^{\mathbb{H}} = \begin{bmatrix} 0 & 0 & 0 & 0 \\[2ex] 0 & 0 & -\left(\dfrac{\partial A_y}{\partial x} - \dfrac{\partial A_x}{\partial y}\right) & \dfrac{\partial A_x}{\partial z} - \dfrac{\partial A_z}{\partial x} \\[2ex] 0 & \dfrac{\partial A_y}{\partial x} - \dfrac{\partial A_x}{\partial y} & 0 & -\left(\dfrac{\partial A_z}{\partial y} - \dfrac{\partial A_y}{\partial z}\right) \\[2ex] 0 & -\left(\dfrac{\partial A_x}{\partial z} - \dfrac{\partial A_z}{\partial x}\right) & \dfrac{\partial A_z}{\partial y} - \dfrac{\partial A_y}{\partial z} & 0 \end{bmatrix} \quad (17.18)$$

This is a chiral field.

Summary:

It would seem that we have the classical electromagnetic field 'leaking' into our 4-dimensional space-time from quaternion space.

Of course, the gauge theory of the standard model has the phase of $SU(2)$ space 'leaking' into our 4-dimensional space-time, but it has no sensible mechanism by which this happens.

In the next two chapters, we will see how well the classical electromagnetic field 'leaks' into our 4-dimensional space-time from quaternion space. In the chapter after the next two chapters, we will return to 'leaky' spaces.

Chapter 18

The Homogeneous Maxwell equations

Both the E-field and the B-field are of a finite group space are fields of the same nature as the potential of that space. Thus, we can differentiate them. Using d to signify the appropriate differential operator, and being careful to keep everything in order (these are non-commutative), we have the B-field differential of the E-field:

$$E = \frac{1}{2}(d\Phi + \Phi d)$$

$$(B \text{ of } E) = \frac{1}{4}(d(d\Phi + \Phi d) - (d\Phi + \Phi d)d)$$

$$= \frac{1}{4}(dd\Phi + d\Phi d - d\Phi d - \Phi dd) \tag{18.1}$$

$$= \frac{1}{4}(dd\Phi - \Phi dd)$$

Similarly, the E-field differential of the B-field is:

$$B = \frac{1}{2}(d\Phi - \Phi d)$$

$$(E \text{ of } B) = \frac{1}{4}(d(d\Phi - \Phi d) + (d\Phi - \Phi d)d)$$

$$= \frac{1}{4}(dd\Phi - d\Phi d + d\Phi d - \Phi dd) \tag{18.2}$$

$$= \frac{1}{4}(dd\Phi - \Phi dd)$$

We see that the B-field differential of the E-field is equal to the E-field differential of the B-field. Further, we see that this is true for finite group space and for any differential operator.

Doing the calculations in detail gives the E-field differential of the B-field of the left-chiral quaternion potential as:

$$\left(\text{E of B}\right)_{L\chi}^{t} = -0 - \frac{\partial B_x}{\partial x} - \frac{\partial B_y}{\partial y} - \frac{\partial B_z}{\partial z}$$

$$\left(\text{E of B}\right)_{L\chi}^{x} = -0 - \frac{\partial B_x}{\partial t}$$

$$\left(\text{E of B}\right)_{L\chi}^{y} = -0 - \frac{\partial By}{\partial t} \qquad (18.3)$$

$$\left(\text{E of B}\right)_{L\chi}^{z} = -0 - \frac{\partial B_z}{\partial t}$$

And the B-field differential of the E-field of the left-chiral quaternion potential as:

$$\left(\text{B of E}\right)_{L\chi}^{t} = 0$$

$$\left(\text{B of E}\right)_{L\chi}^{x} = \frac{\partial E_z}{\partial y} - \frac{\partial E_y}{\partial z}$$

$$\left(\text{B of E}\right)_{L\chi}^{y} = \frac{\partial E_x}{\partial z} - \frac{\partial E_z}{\partial x} \qquad (18.4)$$

$$\left(\text{B of E}\right)_{L\chi}^{z} = \frac{\partial E_y}{\partial x} - \frac{\partial E_x}{\partial y}$$

As we saw above, (18.1) and (18.2), these two double differentials, (18.3) & (18.4) are equal. We have:

$$\frac{\partial B_x}{\partial x} + \frac{\partial B_y}{\partial y} + \frac{\partial B_z}{\partial z} = 0$$

$$-\frac{\partial B_x}{\partial t} = \frac{\partial E_z}{\partial y} - \frac{\partial E_y}{\partial z}$$

$$-\frac{\partial By}{\partial t} = \frac{\partial E_x}{\partial z} - \frac{\partial E_z}{\partial x} \qquad (18.5)$$

$$-\frac{\partial B_z}{\partial t} = \frac{\partial E_y}{\partial x} - \frac{\partial E_x}{\partial y}$$

These, (18.5), are the homogeneous Maxwell equations. They are no more than a non-commutative differential identity. Every finite group space which holds the non-commutative differential has a set of homogeneous Maxwell equations.

Chapter 19

The Inhomogeneous Maxwell Equations

The E-field differential of the E-field is:

$$\left(\text{E of E}\right)^t = \frac{\partial E_t}{\partial t} - \frac{\partial E_x}{\partial x} - \frac{\partial E_y}{\partial y} - \frac{\partial E_z}{\partial z}$$

$$\left(\text{E of E}\right)^x = -\frac{\partial E_t}{\partial x} - \frac{\partial E_x}{\partial t}$$

$$\left(\text{E of E}\right)^y = -\frac{\partial E_t}{\partial y} - \frac{\partial E_y}{\partial t}$$

$$\left(\text{E of E}\right)^z = -\frac{\partial E_t}{\partial z} - \frac{\partial E_z}{\partial t}$$

(19.1)

The B-field differential of the B-field is:

$$\left(\text{B of B}\right)^t = 0$$

$$\left(\text{B of B}\right)^x = \frac{\partial B_z}{\partial y} - \frac{\partial B_y}{\partial z}$$

$$\left(\text{B of B}\right)^y = \frac{\partial B_x}{\partial z} - \frac{\partial B_z}{\partial x}$$

$$\left(\text{B of B}\right)^z = \frac{\partial B_y}{\partial x} - \frac{\partial B_x}{\partial y}$$

(19.2)

The continuity equation requires:

$$\frac{\partial E_t}{\partial t} - \frac{\partial E_x}{\partial x} - \frac{\partial E_y}{\partial y} - \frac{\partial E_z}{\partial z} = 0$$

$$-\frac{\partial E_t}{\partial x} - \frac{\partial E_x}{\partial t} = -\frac{\partial B_z}{\partial y} + \frac{\partial B_y}{\partial z}$$

$$-\frac{\partial E_t}{\partial y} - \frac{\partial E_y}{\partial t} = -\frac{\partial B_x}{\partial z} + \frac{\partial B_z}{\partial x}$$

$$-\frac{\partial E_t}{\partial z} - \frac{\partial E_z}{\partial t} = -\frac{\partial B_y}{\partial x} + \frac{\partial B_x}{\partial y}$$

(19.3)

These are Maxwell's inhomogeneous equations.

Chapter 20

Which Differential Operators do we Want?

If we seek a divergence which matches the distance function, inner product, of the particular finite group space in which we are working, then we must form the differential operator as the adjoint of the potential. Both the adjoint of the potential and the potential have the form of the algebraic matrix form of the finite group space. We can combine (multiply) the adjoint differential operator and the potential because they are both elements of the finite group space.

Multiplication exists within an algebra; it does not properly exist outside of an algebra any more than the sequential combination of permutations exists outside of a finite group. The differential operator acting upon a potential which is an element of a finite group space must be of the form of the finite group space algebraic matrix form, but it is not constrained to be the adjoint of the potential unless we seek a divergence which matches the distance function, inner product, of the finite group space.

Let us just think about the last sentence. We can form a differential operator in any way we like provided it is of the form of the finite group space within which it acts.

If we wish to work within a particular finite group space, then we must use the adjoint form of the differential operator. If we form the differential operator in a way that is different from the adjoint of the potential, then we are 'throwing ourselves' out of the given finite group space.

Above, (13.30), we acted with a differential operator upon a potential within the C_3 3-dimensional finite group space:

$$\begin{bmatrix} \partial a^2 - \partial b \partial c & \partial c^2 - \partial a \partial b & \partial b^2 - \partial a \partial c \\ \partial b^2 - \partial a \partial c & \partial a^2 - \partial b \partial c & \partial c^2 - \partial a \partial b \\ \partial c^2 - \partial a \partial b & \partial b^2 - \partial a \partial c & \partial a^2 - \partial b \partial c \end{bmatrix} \begin{bmatrix} A_a & A_b & A_c \\ A_c & A_a & A_b \\ A_b & A_c & A_a \end{bmatrix}$$

(20.1)

$$= \begin{bmatrix} \dfrac{\partial^2 A_a}{\partial a^2} - \dfrac{\partial^2 A_a}{\partial b \partial c} + \dfrac{\partial^2 A_c}{\partial c^2} - \dfrac{\partial^2 A_c}{\partial a \partial b} + \dfrac{\partial^2 A_b}{\partial b^2} - \dfrac{\partial^2 A_b}{\partial a \partial c} & \sim & \sim \\ & \sim & \\ & \sim & \end{bmatrix} = \begin{bmatrix} Div & \sim & \sim \\ \sim & \sim & \sim \\ \sim & \sim & \sim \end{bmatrix}$$

Because we formed the differential operator as the adjoint of the potential, we arrived at a divergence which matched the distance function, inner product, of the C_3 3-dimensional finite group space. We were working entirely within the given finite group space.

Suppose we had formed the C_3 3-dimensional finite group space differential operator as:

$$
\begin{bmatrix} \partial a & \partial c & \partial b \\ \partial b & \partial a & \partial c \\ \partial c & \partial b & \partial a \end{bmatrix}
\begin{bmatrix} A_a & A_b & A_c \\ A_c & A_a & A_b \\ A_b & A_c & A_a \end{bmatrix} =
\begin{bmatrix}
\dfrac{\partial A_a}{\partial a} + \dfrac{\partial A_b}{\partial b} + \dfrac{\partial A_c}{\partial c} & \dfrac{\partial A_b}{\partial a} + \dfrac{\partial A_c}{\partial b} + \dfrac{\partial A_a}{\partial c} & \dfrac{\partial A_c}{\partial a} + \dfrac{\partial A_a}{\partial b} + \dfrac{\partial A_b}{\partial c} \\[4mm]
\sim & \dfrac{\partial A_a}{\partial a} + \dfrac{\partial A_b}{\partial b} + \dfrac{\partial A_c}{\partial c} & \sim \\[4mm]
\sim & \sim & \dfrac{\partial A_a}{\partial a} + \dfrac{\partial A_b}{\partial b} + \dfrac{\partial A_c}{\partial c}
\end{bmatrix}
$$

$$
= \begin{bmatrix} Div & Curl_B & Curl_C \\ \sim & Div & \sim \\ \sim & \sim & Div \end{bmatrix}
$$

$$(20.2)$$

This differential operator, (20.2), is an entirely valid differential operator within this algebra because it is of the form of this algebra. However, the divergence associated with this differential operator is not of the form of the inner product, distance function, of this algebra. What, if anything, is this divergence, (20.2)? Perhaps this divergence, (20.2), is just squiggles on a bit of paper.

This divergence, (20.2), matches no distance function of any finite group space, and so we do not have this divergence 'leaking' into any other finite group space, but it does match the spatial part of our 4-dimensional space-time. Perhaps this divergence, (20.2), 'leaks' into our 4-dimensional space-time. Well, it might do, but the curls certainly do not match the two term curls that are manifest in our 4-dimensional space-time. We opine that, without an associated set of curls, a divergence is nothing more than squiggles on a piece of paper. This differential operator, (20.2), takes us nowhere.

Other forms of differential operator:

Perhaps we could produce two term curls from the C_3 3-dimensional finite group space by simply setting $\partial c = 0$ in the differential operator. We would have division by zero – that's out of the question. Perhaps we could set a component of the potential to zero. No, the zero component is zero in only a particular co-ordinate system. However we look at it, within a commutative finite group space, because the differential operator and the potential must be of the form of the algebraic matrix form of the finite group space, we will get a set of curls with as many terms as the dimension of the finite group space. Clearly, no differentials of commutative finite group space can be manifest in our 4-dimensional space-time other than the 2-dimensional finite group spaces.

We reiterate that we can differentiate in only finite group spaces because multiplication exists in only finite group spaces.

Looking ahead to 8-dimensions:

Although we do not deal with the 8-dimensional algebras in this book, we ought to make mention of them.

Above, we found that there are at least two valid differential operators within any finite group space. In the C_3 3-dimensional finite group space, these are:

$$\begin{bmatrix} \partial a & \partial c & \partial b \\ \partial b & \partial a & \partial c \\ \partial c & \partial b & \partial a \end{bmatrix} \quad \& \quad \begin{bmatrix} \partial a^2 - \partial b \partial c & \partial c^2 - \partial a \partial b & \partial b^2 - \partial a \partial c \\ \partial b^2 - \partial a \partial c & \partial a^2 - \partial b \partial c & \partial c^2 - \partial a \partial b \\ \partial c^2 - \partial a \partial b & \partial b^2 - \partial a \partial c & \partial a^2 - \partial b \partial c \end{bmatrix} \quad (20.3)$$

It is not hard to construct other C_3 3-dimensional finite group space differential operators, but would they have any application?

When we look at the 8-dimensional algebras, we will be seeking differentials that can be manifest in our 4-dimensional space-time. We will not be interested in the adjoint differential operator.

Chapter 20

Concluding Remarks

Well! We seem to have enjoyed a jolly fine adventure. We began by introducing the finite group spaces. We might think of these as the foundations upon which all of empty space is built. The finite group spaces are very simple and basic constructions that derive directly and quickly from the finite groups and the real numbers. We are adventuring at the foundations of mathematics.

Algebraic structures:

We then considered the algebraic nature of the finite group spaces, and we opted to vary the axioms of a division algebra to include all the finite group spaces rather than include only three of the finite group spaces. It seems most improper that we would allow three representations of finite groups to hold the status of being division algebras while excluding all other finite group representations.

We then examined the nature of the algebraic operations within a division algebra and found that the algebraic operations are really nothing more than:

Addition: The adding of a permutation to the same permutation.

Multiplication: The sequential combination of two permutations.

We then discovered a third algebraic operation which is called the commutator, and the anti-commutator.

Super-positions and our 4-dimensional space-time:

We then introduced the idea of superimposing all members of a set of algebraically isomorphic finite group spaces, and we saw out 4-dimensional space-time emerge as the unique super-position space which can hold any form of rotation – has any form of geometric structure. This is quite a result, but it has been presented elsewhere by your author prior to being presented in this book.

It seemed as if we were digressing from the main theme of this book, but we needed to do this to prepare the road we were about to tread. Super-position is a central part of non-commutative differentiation; we needed to introduce it to the reader.

Inner-products and divergences:

We then went on to consider inner products and divergences and found that each type of space has its own type of inner product (the angle subtended at the origin between two points in the space) and thus its own type of divergence which 'matches' the distance function of that space.

Again, we were preparing the road we were about to tread.

The non-commutative differential:

We were then able to introduce non-commutative differentiation based upon the commutator being a *bona fide* algebraic operation. This led to two differentials which are the E-field and the B-field. Both of these fields are chiral fields, but, when we take the super-position of the different chiralities, we get a non-chiral super-position E-field, and a chiral super-position B-field.

Non-commutative potentials, like the quaternion potential, which we can think of as the $SU(2)$ potential, or the A_3 potentials, which we can think of as the $SO(3,1)$ potential, must be differentiated non-commutatively. Nothing else make sense. Non-commutative potentials will necessarily have two fields associated with them.

Leaky spaces:

We then went on to postulate that fields can leak into spaces from which they were not derived. This is very close to the modern gauge theory view of particle physics.

What is to be done?

We are seeking fields which can be manifest in our 4-dimensional space-time. That means we are seeking fields that have a divergence commensurate with the distance function of our 4-dimensional space-time. We seek a divergence of the form:

$$Div_{\text{4-dim space-time}} = \frac{\partial \phi}{\partial t} - \frac{\partial A_x}{\partial x} - \frac{\partial A_y}{\partial y} - \frac{\partial A_z}{\partial z} \qquad (21.1)$$

With a little thought, and with the divergence of the C_3 finite group space, (13.30), in mind, we might think how lucky we are to have such a simple divergence in our 4-dimensional space-time. We have simple inverses to form our differential operators and a simple divergence.

The fields we seek must also have two-term curls because only two-term curls can be manifest in a 4-dimensional space-time which holds only 2-dimensional rotations.

We have shown that both the quaternion fields and the A_3 fields can be manifest within our 4-dimensional space-time. We believe these fields encompass gravity, classical electromagnetism, and the electro-weak force associated with the electron and the neutrino, but the jury is still out over that. We are missing the strong nuclear force associated with the quarks.

Into 8-dimensional the spaces:

There is much evidence[48] pointing to the one-thousand and twenty-eight 8-dimensional $C_2 \times C_2 \times C_2$ spaces as the origin of the nuclear strong force. The E-fields and the B-fields of these spaces certainly have a divergence that will allow them to 'leak' into our 4-dimensional space-time; in fact, they each have two such divergences – we think electromagnetic force and colour force.

---

[48] Dennis Morris & Sophie Lacson : The Left-handed Spinor

Unfortunately, our present understanding of these 8-dimensional finite group spaces and of the fields within them is still fragmentary and confused. Even more so is our understanding of the 16-dimensional $C_2 \times C_2 \times C_2 \times C_2$ spaces which seem not to 'leak' into our 4-dimensional space-time. It is in this direction that your author's research will continue.

Fare you well:

We hope this book has not shaken your mathematical understanding too violently. Doubtless much which you accepted unquestioningly as established mantra has been pulled out from under you, but such is the way that human understanding deepens.

We hope that, in spite of the large matrices, and many of them, this book has not been a difficult read. You are now at the frontiers of human understanding of empty space. The read was worth it, but it will be a few months before you feel settled with what has been presented to you in this book. Perhaps you will set this book aside for a few months and then re-read it.

It has been a joy adventuring with you.

Dennis Morris

Brotton, Saltburn

January 2018

Other Books by the Same Author

The Naked Spinor – a Rewrite of Clifford Algebra

Spinors exist in Clifford algebras. In this book, we explore the nature of spinors. This book is an excellent introduction to Clifford algebra.

Complex Numbers The Higher Dimensional Forms – Spinor Algebra

In this book, we explore the higher dimensional forms of complex numbers. These higher dimensional forms are connected very closely to spinors.

Upon General Relativity

In this book, we see how 4-dimensional space-time, gravity, and electromagnetism emerge from the spinor algebras. This is an excellent and easy-paced introduction to general relativity.

From Where Comes the Universe

This is a guide for the lay-person to the physics of empty space.

Empty Space is Amazing Stuff – The Special Theory of Relativity

This book deduces the theory of special relativity from the finite groups. It gives a unique insight into the nature of the 2-dimensional space-time of special relativity.

The Nuts and Bolts of Quantum Mechanics

This is a gentle introduction to quantum mechanics for undergraduates.

Quaternions

This book pulls together the often separate properties of the quaternions. Non-commutative differentiation is covered as is non-commutative rotation and non-commutative inner products along with the quaternion trigonometric functions.

The Uniqueness of our Space-time

This book reports the finding that the only two geometric spaces within the finite groups are the two spaces that together form our universe. This is a startling finding. The nature of geometric space is explained alongside the nature of division algebra space, spinor space. This book is a catalogue of the higher dimensional complex numbers up to dimension fifteen.

Lie Groups and Lie Algebras

This book presents Lie theory from a diametrically different perspective to the usual presentation. This makes the subject much more intuitively obvious and easier to learn. Included is perhaps the clearest and simplest presentation of the true nature of the Lie group $SU(2)$ ever presented.

The Physics of Empty Space

This book presents a comprehensive understanding of empty space. The presence of 2-dimensional rotations in our 4-dimensional space-time is explained. Also included is a very gentle introduction to non-commutative differentiation. Classical electromagetism is deduced from the quaternions.

The Electron

This book presents the quantum field theory view of the electron and the neutrino. This view is radically different from the classical view of the electron presented in most schools and colleges. This book gives a very clear exposition of the Dirac equation including the quaternion rewrite of the Dirac equation. This is an excellent introduction to particle physics for students prior to university, during university and after university courses in physics.

The Quaternion Dirac Equation

This small book (only 40 pages) presents the quaternion form of the Dirac equation. The neutrino mass problem is solved and we gain an explanation of why neutrinos are left-chiral. Much of the material in this book is drawn from 'The Electron'; this material is presented concisely and inexpensively for students already familiar with QFT.

An Essay on the Nature of Space-time

This small and inexpensive volume presents a view of the nature of empty space without the detailed mathematics. The expanding universe and dark energy is discussed.

Elementary Calculus from an Advanced Standpoint

This book rewrite the calculus of the complex numbers in a way that covers all division algebras and makes all continuous complex functions differentiable and integrable. Non-commutative differentiation is covered. Gauge covariant differentiation is covered as is the covariant derivative of general relativity.

Even Mathematicians and Physicists make Mistakes

This book points out what seems to be several important errors of modern physics and modern mathematics. Errors like the misunderstanding of rotation, the failure to teach the higher dimensional complex numbers in most universities, and the mathematical inconsistency of the Dirac equation and some casual errors are discussed. These errors are set in their historical circumstances and there is discussion about why they happened and the consequences of their happening. There is also an interesting chapter on the nature of mathematical proof within our society, and several famous proofs are discussed (without the details).

Finite Groups – A Simple Introduction

This book introduces the reader to finite group theory. Many introductory books on finite groups bury the reader in geometrical examples or in other types of groups and lose the central nature of a finite group. This book sticks firmly with the permutation nature of finite groups and elucidates that nature by the extensive use of permutation matrices. Permutation matrices simplify the subject considerably. This book is probably unique in its use of permutation matrices and therefore unique in its simplicity.

Index

www.ingramcontent.com/pod-product-compliance
Lightning Source LLC
Chambersburg PA
CBHW081729220526
45468CB00008B/2030